CANAL SYSTEMS AUTOMATION MANUAL

Volume 1

A Water Resources Technical Publication

A guide to the use of engineering technology pertaining to selecting automatic control schemes for canals conveying water to irrigable lands.

Scientific Publishers

U.S. DEPARTMENT OF THE INTERIOR
Bureau of Reclamation

CANAL SYSTEMS AUTOMATION MANUAL

VOLUME 1

By

C.P. Buyalski
D.G. Ehler
H.T. Falvey
D.C. Rogers
E.A. Serfozo

Project Management by
D.L. King

First Edition, 1991

Hydraulics Branch
Research and Laboratory Services Division

Electrical Branch
Electrical and Mechanical Engineering Division

Water Conveyance Branch
Civil Engineering Division

SCIENTIFIC PUBLISHERS (INDIA)
P.O. BOX 91 JODHPUR

ISBN : 81-7233-045-6

First reprint in India, 2000

by
PAWAN KUMAR
Scinetific Publishers (India)
5-A, New Pali Road, P.O. Box 91
JODHPUR -342 001 (India)
Fax : +91-291-512580
E-mail : artrans@jp1.dot.net.in

The Book is published with the permission of
the United States Department of the Interior
Bureau of Reclamation, Denver, Colorado, 1987

Printed in India

PREFACE

Since its inception in 1902, one of the Bureau of Reclamation's prime objectives has been the supply of irrigation water to arid lands. This involves the construction of dams to store water and of canals to distribute the water. Like the artisans of a millennium, the Bureau has sought to improve the methods of water distribution. The earliest designs of canals primarily were based upon the maximum flow condition. This design does not provide the requisite flexibility to operate a canal system efficiently. To improve operations, the concepts of automation were conceived (circa 1950). The first canal automation attempts were crude, but they were immediately successful. Advances in the operation of canals through the use of automation paralleled the development of electronics in industry. Presently, some Bureau canals and dams are being controlled by the most sophisticated computer equipment and programs that are available.

The first research program in canal automation began in 1966 by the Bureau's Mid-Pacific Region. The results of these studies were applied to canals in California. The need for a much broader research program soon became apparent—one which would apply to all the Bureau's regions. Therefore, in 1971, the Bureau's Engineering and Research Center (now the Denver Office) formed a water systems automation team in Denver, Colorado. The team members consisted of engineers from the planning, design, and research disciplines. The purpose of the team was to coordinate and direct a program of canal automation research development and application. Annual meetings were held with representatives from each region to ensure that the program conformed to regional needs.

A positive aspect of the interdisciplinary team effort was in strengthening of the canal automation capabilities and in developing technical staff in many of the regions and in the design groups at the Denver Office. As regional and design staff's expertise developed, the need for an interdisciplinary team at Denver gradually decreased. However, the division chiefs (at Denver Office) recognized a need for consolidating the automation knowledge — gained by the interdisciplinary studies — within a manual.

In 1986, a Water Systems Automation Management Team, which consisted of branch chiefs, was formed at the Denver Office to help prepare a manual that incorporated the latest Bureau experience in water systems automation designs. The management team created an author team consisting of five engineers—three from the Research Division and two from the Design Division—to develop the manual. All the engineers were active in the interdisciplinary effort. Additionally, a representative from the Water Operations and Maintenance Branch (now Facilities Engineering Branch) and a technical editor were appointed to the author team to assist the authors in its preparation (scheduling, coordination, editing, and illustration).

The manual is primarily for:

- Managers and operation engineers of water conveyance systems who are interested in implementing advanced water operations techniques for more effective and efficient operations.
- Designers, researchers, and operation engineers who have technical oversight responsibilities in a staff role for review of project operations.

In September 1986, the author team began its endeavor; subsequently, it was decided to publish the manual in three volumes; it would address the needs of the different audiences. The volumes would progress from fundamentals of canal automation to complex control theory applications. The level of detail in the first volume would be sufficient for a planner or system manager to understand the various control concepts. The second volume is to be oriented toward planning and design engineers. It would delve deeper into control theory, illustrating the concepts with equations, methods of simulation, and hardware technology. However, the detail will not be at a level that would allow one to perform transient analysis of a system or to develop values for control parameters. The third volume is intended to focus on selected topics in research and development that had not been implemented on Bureau projects. It would be of interest to research and development specialists.

Previously published information related to the details of automated canal systems, as well as the availability of numerical models, would be identified in the manual. Part of the manual consists of standardized definitions of specific automation terms. A glossary expands on the usage most prevalent in the Bureau of Reclamation.

Volume I is by the following team:

Clark P. Buyalski (retired) ..... Hydraulics Branch
David G. Ehler .... Hydraulics Branch
Henry T. Falvey (retired) ...... Hydraulics Branch
David C. Rogers ...... Water Conveyance Branch
Edward A. Serfozo .... Electrical Branch
David L. King ...... Facilities Engineering Branch
Richard N.Walters .. Publications & Records Branch

CONTENTS

CONTENTS — Continued

CONTENTS — Continued

CONTENTS — Continued

FIGURES

CONTENTS — Continued

CHAPTER 1

INTRODUCTION

SCOPE OF THE MANUAL

1-1. General

Practical aspects of canal operation, control, and automation are presented in this manual. The manner in which subjects are organized begins with fundamentals and basic concepts, continues with more specific details, and finishes with the more complex aspects of control theory analysis. Available and successful control techniques discussed in the manual are presented from the method standpoint. (Specific equipment is not discussed in volume 1.) Thus, the control methods will not become outmoded as equipment technology advances.

1-2. Purpose

This manual has two principal purposes. One is to consolidate Bureau of Reclamation (the Bureau) experience in canal system automation into a single source reference. The other is to provide information and guidelines for managers, operators, and engineers to select successful and practical canal system automation schemes.

Those who have daily responsibility for delivering water from its source to the point of diversion should find the manual quite useful. The intent is that the manual provides:

- Definitions of terminology
- Fundamentals of operation and control
- Technical and design considerations
- Important research and development topics
- Practical examples of automated canal systems implemented at selected Bureau projects

1-3. Organization

The manual is divided into three parts that address three levels of readership:

Volume 1—Water users and operators
Volume 2—Planning and design engineers
Volume 3—Research and development specialists

To achieve this goal, the manual progresses from fundamentals to complex control theory applications. This arrangement allows the reader to concentrate on that area which is of most interest. However, the authors recommend that the reader review other parts of the manual to expand their familiarity with canal operational requirements and objectives. Selection of a practical control system can be made after expertise is acquired.

Volume 1 describes the fundamentals of operation and control. Managers, operators, and water users of canal systems will find it to be of particular interest. Chapter 1 introduces the manual and subject of canal systems automation. Chapter 2 details the fundamental principles of operating a canal system including basic open channel hydraulics and delivery concepts, and the operational and structural constraints imposed by a canal system. Chapter 3 identifies basic control theory and the methods that are available for regulating canal check gate structures. Chapter 3 also discusses basic control algorithms. In Chapter 4, sufficient information is provided to enable a watermaster and engineer to evaluate and select a practical control method that would upgrade the operation of an existing canal system.

Volume 2 will present additional technical aspects of canal operation and control. The operational requirements and the various control methods shall be considered in detail. Planning and design engineers should benefit from this part of the manual as they will obtain a better understanding of the technical requirements for implementing the selected control system. Chapter 5 is intended to emphasize control theory and applications; that is, it involves control responses, specific applications of various control methods, and the more sophisticated state-estimation (prediction) algorithms. Chapter 6 is projected to be an important section for the design engineer of canal systems. Fuller treatment will be devoted to the design process and considerations involved with operation, physical features, and actual control of the canal system. Simulation by modeling the canal operation is important in certain phases of control theory application as chapter 7 unfolds. Chapters 8 through 11 will present hardware technology. These chapters will include descriptions of communication systems, instrumentation, electrical and mechanical systems, and canal automation equipment.

Volume 3 is intended to present selected topics in research and development. Important information regarding complex control theory applications would be considered. Research and development specialists should find the information useful. The require ment for establishing water motion stability while applying closed loop feedback control systems to automatically regulate water control structures will be included in volume 3. Potential control concepts that have been studied, which are available, but have not been applied to actual Bureau canal systems will be included.

An appendix that includes examples of canal automation projects currently in operation is planned for volume 3. They are specific examples of control methods on existing canal systems, and should aid one in selecting an appropriate control system.

A glossary of special terms and phrases used in canal system automation is provided here. Many terms and phrases used in the manual sometimes have special meanings as applied to canal system automation.

BASIC TERMINOLOGY

Good communication between those involved in selecting, planning, designing, and implementing a successful canal system automation project is an essential ingredient in the design process. A common understanding of terms and phrases that describe canal system automation technology is necessary to ensure a successful project. The following definitions introduce and define the essential terminology used in the discussion of canal system automation projects.

1-4. Canal Automation

Automation is a common engineering term. It is defined here as:

> A procedure or method used to regulate a system by mechanical or electronic equipment that takes the place of human observation, effort, and decision; the condition of being automatically controlled.

Thus, canal automation is defined as:

> The implementation of a control system that upgrades the conventional method of canal system operation.

Several terms in the preceding definition require further refinement:

Control system—An arrangement of electronic, electrical, and mechanical components that commands or directs the regulation of a canal system.

Upgrades—Provides a better match between the canal system delivery capabilities and the water users' demands. As a result, improved response and system efficiency is achieved beyond what could be accomplished by the conventional method of operation.

Conventional method—Canal system control onsite by operations personnel (ditchrider and watermaster). Labor saving devices and machinery may be used to assist in the control of the canal facilities. The ditchrider is responsible for controlling the canal system onsite based upon the flow schedule established by the watermaster at the central headquarters. The watermaster is responsible for overall operation.

Canal system operation—Water transfer from its source to points of diversion for irrigation, municipal and industrial, fish and wildlife, and drainage purposes.

The definition of canal automation states that the operation of the canal system is improved or upgraded by installing a control system. Upgrading should not be confused with updating. Even though a control system can be updated by replacing it with new equipment using the latest technology available, the actual operation may or may not be upgraded.

1-5. Water Conveyance Systems

The water conveyance system (water distribution system) is one of the important elements of the

overall water project. Basically, it is constructed to convey water from one place to another. Several objectives for conveying water are:

- Irrigate agricultural land
- M&I (municipal and industrial) expansion
- Fish and wildlife enhancement
- Increase project water supply
- Decrease flood damage
- Power generation

a. **Classification.**—The objectives of a water conveyance system determine their general classifications. The classifications are:

Delivery system—Conveys water from a single source such as a storage reservoir to a number of individual points of use. The delivery system is a common classification. It is associated with irrigation, M&I, and fish and wildlife canal systems.

Collector system—Conveys water from several individual sources such as ground-water wells and drains and surface inlet drains for rain storm and snowmelt runoff to a single point of diversion. The collector system is associated with projects that increase water supply and decrease flood damage.

Connector system—Conveys water from a single source to a different location—typically without intermediate collection or diversions. The connector system is associated with intakes to pumping plants and/or powerplants and with regulation reservoirs.

Figure 1-1 is a schematic of a main canal system with the three types of classification. A delivery

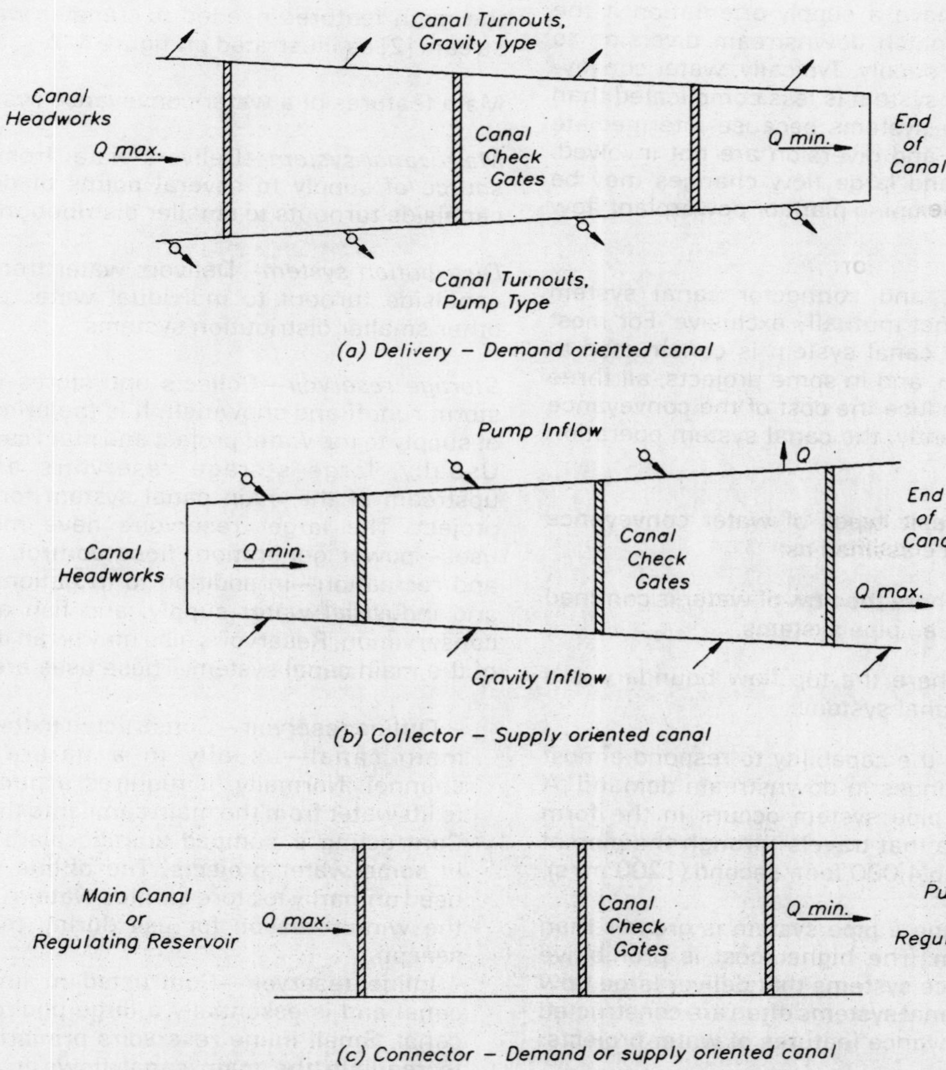

Figure 1-1. — Schematic plans of three main canal systems.

system that is demand-oriented is represented on figure 1-1a. It must convey the available water supply from the canal headworks to downstream turnouts, and must match variable demands for any number of diversion points. Downstream demands establish the canal system flow requirements or schedule.

Conversely, the collector system is supply-oriented as shown on figure 1-1b. It must convey a variable water supply from any number of sources to a downstream single point of diversion. Typically, a collector system does not have canalside diversions (turnouts). Upstream water inflow sources establish the canal system flow requirements or schedule.

The connector system noted on figure 1-1c can be either demand- or supply-oriented. The system has a demand-orientation if the objective is to match the upstream water supply to a variable downstream demand. It would have a supply-orientation if the objective were to match downstream diversion to the upstream water supply. Typically, water conveyance in a connector system is less complicated than delivery or collector systems because intermediate points of collection and diversion are not involved. However, sudden and large flow changes may be required to match pumping plant or powerplant flow fluctuations.

Delivery, collector, and connector canal system classifications are not mutually exclusive. For most water projects, the canal system is constructed to include at least two, and in some projects, all three classifications to reduce the cost of the conveyance features. Consequently, the canal system operation can be complex.

b. Types.—The basic types of water conveyance systems [1][1] can be classified as:

Closed conduit—Where the flow of water is confined on all boundaries; i.e., pipe systems.

Open channel—Where the top flow boundary is a free surface; i.e., canal systems.

Pipe systems have the capability to respond almost immediately to changes in downstream demand. A flow change in a pipe system occurs in the form of a pressure wave that travels through the pipe at a high speed—up to 4,000 feet/second (1200 m/s).

Cost for constructing a pipe system is greater than for a canal system. The higher cost is prohibitive for most conveyance systems that deliver large flow rates. Therefore, canal systems often are constructed as the major conveyance features of water projects.

Flow in a canal system is more difficult to regulate than in a pipe system. Flow changes in canals occur in the form of translatory wave fronts. The translatory wave travels at a relatively slow speed, about 200 times slower than the pressure wave in a pipe system. Typical canal translatory wave speed range is 10 to 20 feet per second (3 to 6 m/s); therefore, a canal system responds slower to changes in downstream demand than a pipe system does.

Primarily, this manual deals with control of canal systems. A discussion of automating a pipe system would require an equivalent treatise and is beyond the scope of this manual.

1-6. Main Features of a Water Conveyance System

The water conveyance system can be divided into the main features needed to transfer water downstream [2] as illustrated on figure 1-2.

Main features of a water conveyance system are:

Main canal system—Delivers water from a primary source of supply to several points of diversion or canalside turnouts to smaller distribution systems.

Distribution system—Delivers water from the main canalside turnout to individual water users or to other smaller distribution systems.

Storage reservoir—Collects and stores water from storm runoff and snowmelt. It is the primary source of supply to the water project and main canal system. Usually, large storage reservoirs are located upstream of the main canal system for the water project. The larger reservoirs have multipurpose uses—power generation, flood control, navigation, and recreation—in addition to irrigation, municipal and industrial water supply, and fish and wildlife conservation. Reservoirs also may be an integral part of the main canal system. These uses are:

- Offline reservoir—Constructed to the side of the main canal—usually in a natural drainage channel. Normally, it requires a pumping plant to lift water from the main canal into the reservoir. Contructing a pumped storage plant is feasible in some water projects. The offline reservoir is used primarily to store surplus water runoff during the winter season for use during the irrigation season.
- Inline reservoir—Contructed in line with the canal and is essentially a large pool of the main canal. Small inline reservoirs primarily are used to regulate the main canal flows to maintain a balanced operation. However, large inline reservoirs also may be used for carryover storage.

[1] Numbers in brackets refer to bibliography.

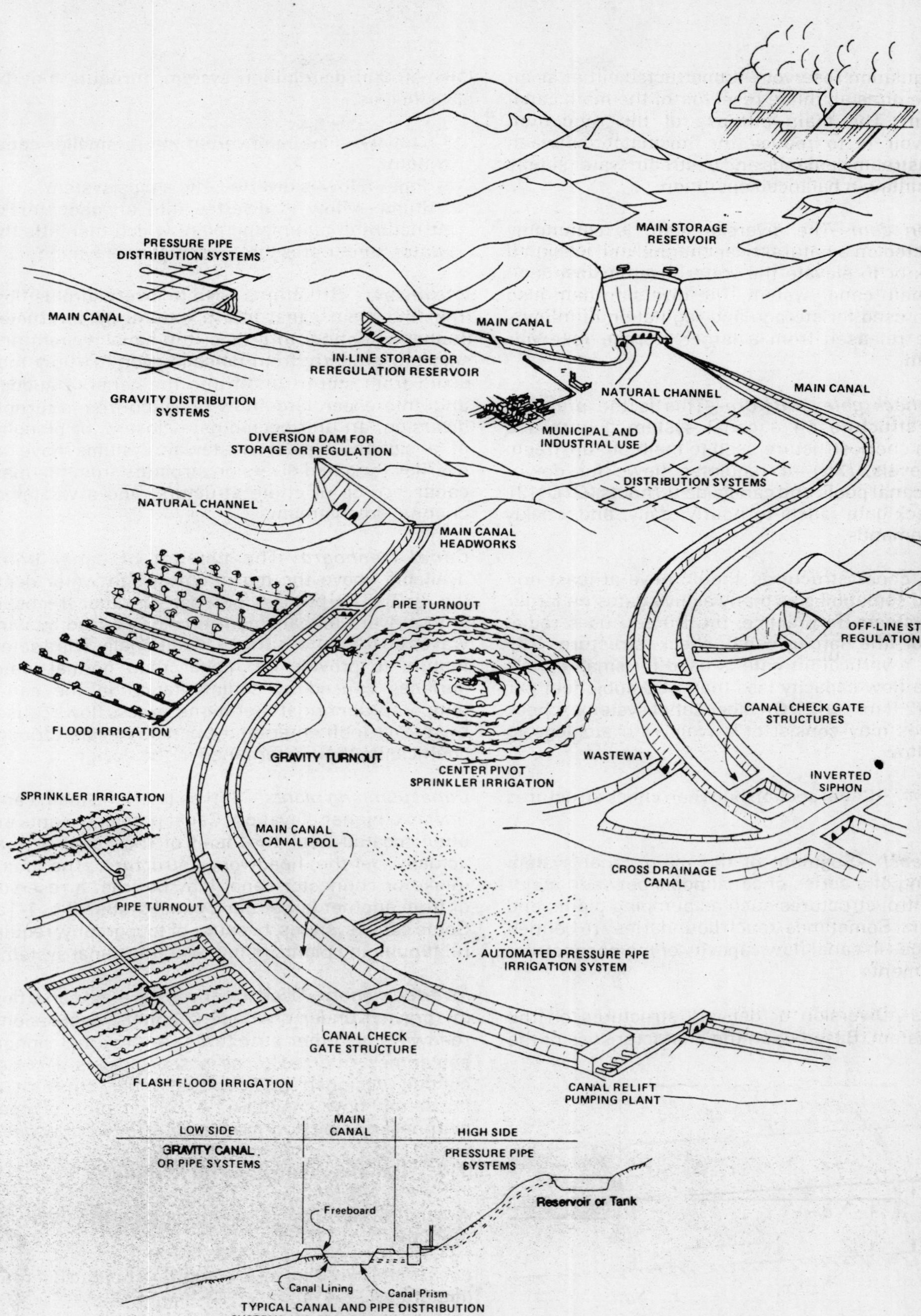

Figure 1-2. — Main features of a water conveyance system.

- Regulation reservoir—Constructed either as an offline or as an inline reservoir of the main canal system. The main purpose of the regulation reservoir is to reduce the mismatch between downstream demands and upstream water supply to maintain a balanced operation.

Diversion dam—The diversion dam is commonly constructed on a natural river channel and designed to check or to elevate the water level for diversion into a main canal system. The diversion dam also could be used for storage and regulation of inflows that are released from a larger storage reservoir upstream.

Canal check gate structure—This is the principal control structure on a canal system. Figure 1-3 shows a check structure used to maintain upstream water levels, (*YU*)—it regulates flows for downstream canal pools and canalside turnouts (*QTO*) [3]. The check gate reacts to hourly, daily, and weekly water demands.

Usually, check structures should have at least one gate and sometimes as many as five gates on larger canal systems. In practice, the Bureau uses radial gates for the larger canal check structures [3]. Usually, a vertical lift gate is used for small canals having a flow capacity less than 100 cubic feet per second (2.8 m^3/s) [2]. On older canal systems, check structures may consist of a weir with stoplogs to control flow.

Canal pool—Canal section between check structures (fig. 1-3).

Canal reach—Segment of the main canal system consisting of a series of canal pools between major flow control structures such as pumping plants and reservoirs. Sometimes, reach boundaries are defined by changes in canal flow capacity or major construction segments.

Turnouts—Diversion or delivery structures on the canal system. Based upon the characteristics of the downstream distribution system, turnouts may be classified as:

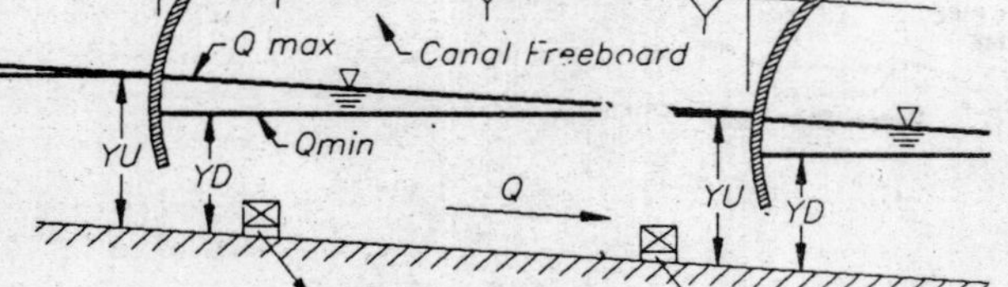

Figure 1-3. — Profile of main canal system depicting canal check gate structures and canalside turnouts.

- Gravity—Flow is diverted into a smaller canal system.
- Pipe—Flow is diverted into a pipe system.
- Pump—Flow is diverted into an open intake structure of a pumping plant which then lifts the water, under pressure, to a higher elevation.

Wasteways—Structures used to divert surplus flow from the main canal into a natural or contructed drainage channel to protect the canal system and adjacent property from damage. Surplus flow may result from storm runoff into the canal, a sudden and unforeseen large flow rejection from a turnout diversion, an unexpected gate closure, or plugging of a siphon. Typical wasteway systems have an overflow or gated diversion structure from the main canal, a drop or chute structure, and a wasteway channel downstream.

Canal freeboard—The amount of canal lining available above the maximum design water depth (fig. 1-3). The primary purpose of canal freeboard is to prevent embankment erosion caused by wind waves on the water surface. Additionally, it is a safety feature to provide for higher than normal water surfaces caused by sedimentation in the canal, temporary abnormal operations, excess flows caused by storm runoff entering the canal, or higher friction coefficients than determined in the design.

Canal pumping plants—Serves to lift the main canal flow to a higher elevation. Canal pumping plants are often located near the head of the canal or may actually be the headworks structure. The pump intake or connector canal may be from a reservoir or from another larger main canal system (fig. 1-1c). Some canal systems, because of topography, require relift pumping plants in line with the canal system.

Inverted siphon—Used to convey the main canal flow (by gravity) under drainage channels, depressions, roadways, or other structures. An inverted siphon, sometimes referred to as a sag pipe, is a closed conduit designed to flow full under pressure at maximum flow condition. A combination of open channel and pipe flow may occur at lesser flow rates.

1-7. Canal System Control Methods

A canal system can be controlled by the following methods:

Local manual—Control is onsite by a human operator (ditchrider).

Local automatic—Control is onsite by control equipment without human intervention.

Supervisory—Control is from central headquarters with different levels of participation by the watermaster.

Combined—A combination of the above methods is used.

Each canal system control method has its own characteristics and advantages. The selection of the method that will upgrade a canal operation does not require knowledge of the associated control equipment details. However, selection is made easier when each basic method of control is clearly understood. Control fundamentals and selection of an appropriate control method are discussed more extensively in chapters 3 and 4. The following briefly discusses each method.

a. **Local manual control**.—Local manual control is the conventional method of controlling a canal system. Labor saving devices and machines, such as a gate hoist motor, may be used onsite to assist the ditchrider when adjusting canal check gate openings. The procedure basically involves:

- An order by the water user
- A change (increase or decrease) in the input of water at the head of the canal to meet that order
- The adjustments by the ditchrider at the check gate and canalside turnout structures when the wave front arrives

An experienced ditchrider can anticipate water arrival time within acceptable limits. This type of operation is greatly personalized (subjective). On a large canal, many people are involved. The ditchrider (in the field) must report—by mobile radio or telephone—canal data such as water levels and gate positions to the watermaster. The watermaster (at the central headquarters) needs to know these data to implement the schedule upon the complete system. Team effort is involved; all members must use judgment in their respective area of responsibility. Successful operation depends in large measure on the team's skill.

b. **Local automatic control**.—Local automatic control allows unattended canal operation achieved through an arrangement of mechanical, electrical, and electronic components located onsite. The equipment monitors water depths and gate positions. The sensed information is interpreted by specially developed equations referred to as control algorithms. The algorithms are designed to calculate adjustments required to position the canal check gates and to satisfy the actual operating needs.

For the development of this method, one must be capable of blending technical knowledge and practical experience into the design of the local automatic control system. Control algorithms and control parameters are developed by engineers having knowledge of control theory. Their procedures depend upon the experience of the canal watermaster and ditchriders. Therefore, using a combination of control engineering and operators' experience is essential for this method of operation.

Required onsite human input is reduced to periodically monitoring the canal operation. Automatic control equipment does not require onsite manual adjustments by the ditchrider to achieve proper canal operation. However, an alarm system is required to notify central headquarters of abnormal conditions such as control equipment failure, high or low water levels, local power outages, and communication channel failures. Therefore, a communication channel is necessary between each check gate controller and the headquarters to provide alarm information. The watermaster (at headquarters) monitors the alarm information and promptly initiates corrective action based upon the abnormal condition.

c. **Supervisory control**.—Supervisory control is the operation of the canal system by the watermaster from a central location referred to as the headquarters or master station. The master station equipment performs the function of collecting data from remote sites on the canal system, analyzing the data, and presenting the data in a suitable format for further action by the watermaster. Each remote site, such as the canal check gate structure, requires a remote terminal unit (RTU). The RTU monitors data such as water levels, gate positions, and equipment status, and then transmits these data to the master station. The RTU also controls the remote site; it adjusts the gate position based upon the watermaster's instructions received from the master station. Therefore, this type of operation requires a two-way communication system between each RTU at the remote sites and the master station.

Direct control of the canal system is the watermaster's responsibility at the headquarters. Effectively, many ditchrider's duties are transferred from the field location to a centralized location. Supervisory control enables all required information from the entire canal system to be centralized at one location. Changes in one portion of the system can be recognized promptly and appropriate control action initiated. Flow can be adjusted simultaneously at many canal check gate structures resulting in a mass transfer of water in the shortest time possible—providing an improved or upgraded operation.

d. **Combined control systems**.—Local manual, local automatic, and supervisory control systems are not mutually exclusive. In many water projects, the operation involves using a combination of two or perhaps all three methods. The relative merits of the three types of control depend on the circumstances of use. For example, a gravity type turnout may require a local automatic controller to automatically maintain a constant delivery rate if the main canal water level changes frequently. The ditchrider is not always available to visit the site frequently enough to make the necessary gate opening adjustments to compensate for the water level variation and maintain a constant delivery rate.

A turnout local automatic controller installation upgrades a turnout gate operation. However, the main canal system supplying water to the turnout may still operate in a conventional manner; i.e., by the local manual control method. Instead of changing the turnout gate position for a new delivery rate, the ditchrider changes the local automatic controller setpoint position. Hence, the operation combines local manual and local automatic control methods and the canal operation has been upgraded.

Operation can be upgraded further by installing local automatic controllers at the main canal check gate structures. However, the turnout local automatic controller still requires daily setpoint changes. This could be accomplished by supervisory manual control. The watermaster (at the master station) transmits the new setpoint to the appropriate turnout local automatic controller at the proper time—relieving the ditchrider from this task. This example combines local automatic and supervisory manual control. The master station equipment could be programmed to determine the proper time for the setpoint change based upon the current demand schedule. Then, the new setpoint is transmitted automatically to the proper turnout local automatic controller at the appropriate time. Local automatic control is combined with supervisory computer-directed control and frees the watermaster from normal daily water scheduling tasks.

CANAL SYSTEM AUTOMATION FUNDAMENTALS

Generally, canal systems are constructed to convey water from a source to downstream diversion points in the most efficient and economical manner possible using the available resources. Usually, automation of canal systems upgrades and enhances the overall canal operations for increased crop yields to the farmers.

1-8. Purpose and Benefits

The primary purpose of canal system automation is to upgrade a canal operation. Canal operation is enhanced by installing a practical and modern control system to improve the water transfer efficiency. Many limitations inherent in the conventional method of operation can be overcome. Local automatic control, supervisory control, or combination control can provide the following advantages compared to the conventional method:

- Rapid transfer of water (small or large quantities) with frequent changes
- Achieve operational flexibility when simultaneous information on the entire system is known
- Adjustments in canal flow can be made to accommodate variances in canalside turnout diversions on a daily or an hourly basis
- Immediate response to sudden and unannounced variances in canalside diversion or to storm runoff flooding into the canal.

Advantages inherent in canal automation hold interest for canal operators. A control system can be designed to ensure coupling of the canalside turnout diversions and the canal inlet supply. An automated canal system can respond to abrupt and unscheduled changes in turnout diversion and immediately initiate corrective action. The action reduces the necessity for excess flow and minimizes the occurrences of shortages in the lower canal pools. A more optimum transfer of water can be attained with an automated canal.

The objective of canal system automation is to realize the benefits associated with implementing a practical control system. Some benefits are:

- Better service to the water user
- Efficient transfer of water
- Reduction in operating costs
- Minimizes diversions into wasteways

The water project should provide operating benefits even if reductions in tangible operating cost only offsets the cost of implementing an automated control system. Intangible benefits associated with improved water service and water transfer are significant. However, the monetary value of intangible benefits is not easily assessed.

Water users are the main beneficiaries. Water users receive a scheduled quantity of water at the time specified without the limitations imposed by the conventional method of operation. Water users should realize economic benefits and better service. By providing the water user with improved service the watermaster would certainly receive fewer complaints throughout the growing season.

Additionally, efficient transfer of water could result in an economic benefit if the canal system has a history of wasting water. Projects having to schedule power for relift pumping could realize a substantial economic savings by not pumping surplus flow.

Significant savings could result from lower maintenance costs associated with canal lining and embankment failures. Efficient water transfer requires maintaining near steady-state water levels in the canal. The frequency of excessive pressure changes on canal lining and recurrence of water levels above the canal freeboard can be greatly reduced.

Reduction in operation and maintenance (O&M) costs associated with installation of a control system constitutes the main tangible economic savings. The number of ditchriders needed to operate the canal system after the control system is placed in operation can be reduced. The additional ditchriders needed to provide conventional service to the water users should be included in the benefit/cost analysis. However, even with canal automation, some ditchrider coverage of the canal system is needed.

Automation creates a need for technical expertise to maintain computer software and electronic equipment. Technical skills are required to apply general principles of control system development and implementation and to repair problems. Development and implementation work would probably be accomplished as part of the initial specifications contract and be considered as a capital cost of the project. However, skilled technicians are required to perform maintenance.

The decision to implement an automated control system should not be justified solely on tangible economic benefits. Canal operations vary from one canal to the other—depending on the type of inlet and outlet facilities, the flow capacity, and the particular combination of service demands. Before a cost/benefit analysis can be accomplished, the operational requirements and specific needs to automate the distribution system have to be identified. Proper identification of the requirements and needs leads to either the selection of an appropriate automation system or continued operation using the existing method. Additional information on the decision process is in chapter 4.

1-9. Canal Operation and Control Concepts

The operation concept, established by the canal system operating criteria, determines the flow schedule. The control concept determines how the canal control structures are adjusted to satisfy the operation concept. Both operation and control concepts depend on upstream or downstream conditions.

a. Canal operation concepts.—The operation concepts used to determine the schedule of a canal system are defined by whether the main canal water schedule (flow) is based upon the downstream canalside turnout demand (or scheduled deliveries) or based upon the upstream water supply.

(1) Downstream operation.—Downstream water demand or scheduled delivery determines the canal system flow schedule. Downstream operation concept applies to canal systems that are primarily demand-oriented and usually is associated with delivery systems (see sec. 1-5 and fig. 1-1a).

(2) Upstream operation.—Upstream water supply source or inflow determines the canal system flow schedule. Upstream operation concept applies to canal systems that are primarily supply-oriented and usually the concept is associated with collector systems (see sec. 1-5 and fig. 1-1b).

b. Canal control concepts.—The concepts used to control the canal system are defined by location of information needed for control relative to the control structure, as diagramed on figure 1-4.

(1) Downstream control.—Control structure adjustments are based upon information from downstream (fig. 1-4a). The required information could be measured by a sensor located downstream or be based upon the downstream water schedule established by the watermaster. Downstream control transfers the downstream canalside turnout demands to the upstream water supply source (or canal headworks) and is compatible with the downstream operation concept.

(2) Upstream control.—Control structure adjustments are based upon information from upstream (fig. 1-4b). The required information could be measured by a sensor located upstream or be based upon the upstream water schedule established by the watermaster. Upstream control transfers the upstream water supply (or inflow) downstream to points of diversion or to the end of the canal and is compatible with the upstream operation concept.

c. Concept compatibility.—The operation and control concepts must be compatible as depicted on figure 1-5. The compatible schemes are:

- Downstream operation concept with downstream control concept
- Upstream operation concept with upstream control concept

It is impossible to operate supply-oriented canal system having an upstream operation concept using the downstream control concept. Additionally, it is

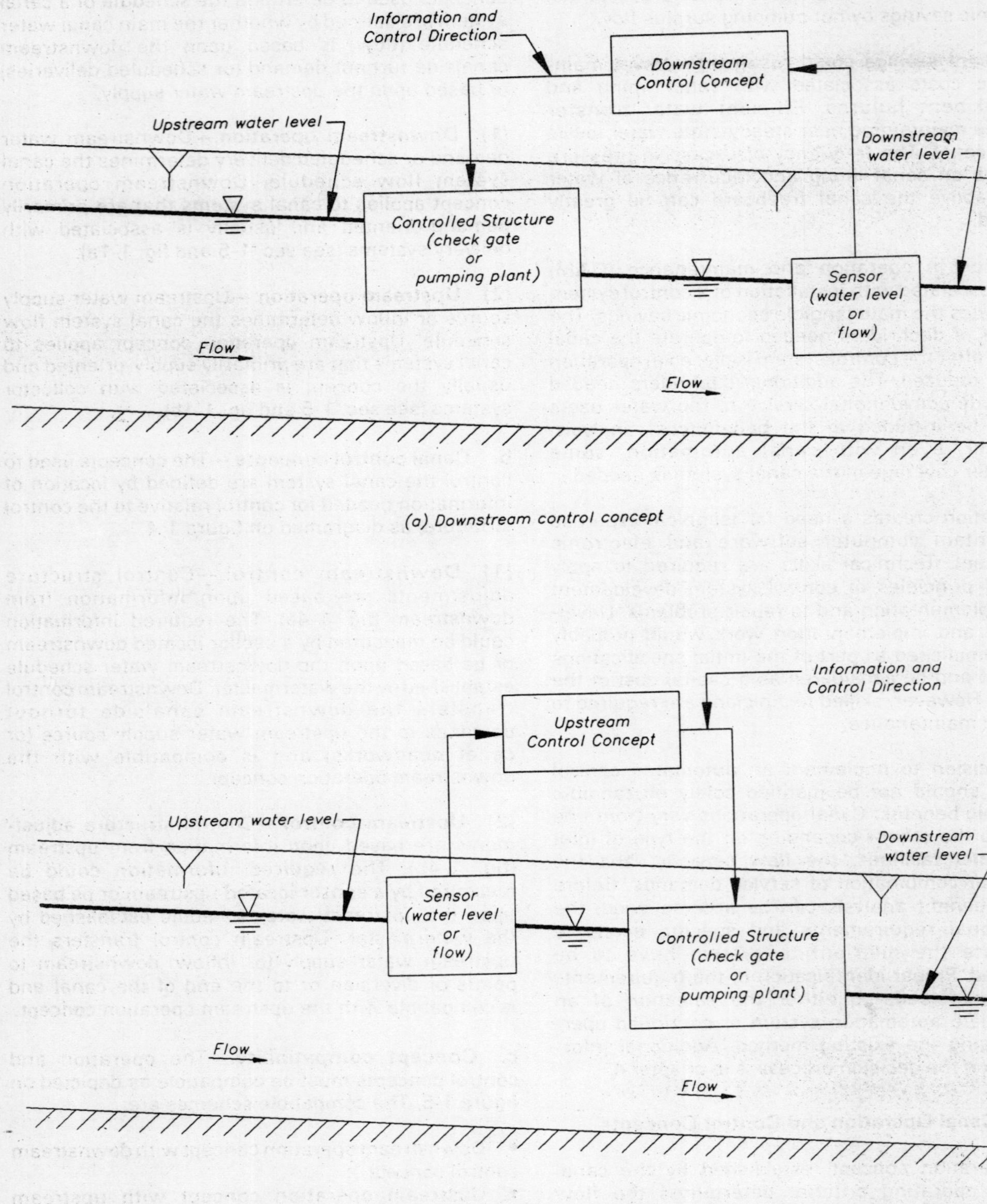

(a) Downstream control concept

(b) Upstream control concept

Figure 1-4. — Schematic of downstream and upstream control concepts.

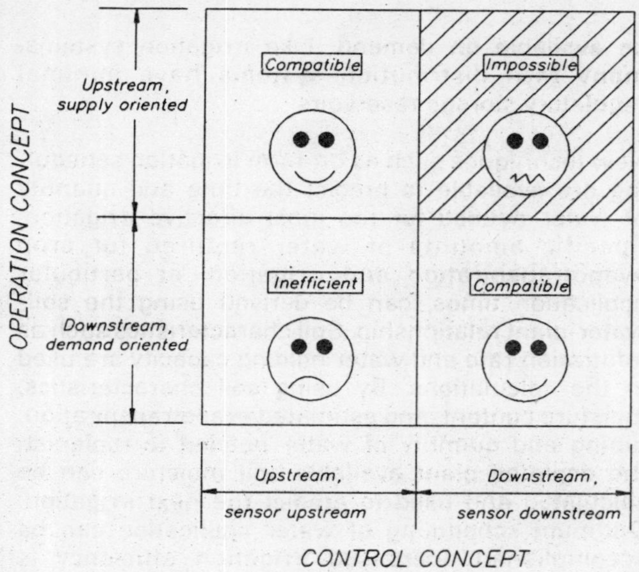

Figure 1-5. — Operation and control (upstream and down stream) concept compatibility.

impossible to apply both the downstream and upstream concepts (operation or control) to a canal system simultanously. For most demand-oriented canal systems, using the downstream operation concept would be inefficient to apply the upstream control concept.

It is possible to operate certain canal systems using either the downstream or upstream operation concept if the basic objective of the canal system changes. For example, a canal system can be designed primarily to provide water to farmers during the irrigation season—a demand-oriented operation. Whereas, the canal system can convey surplus flood flows which occur during the off-irrigation season into an offline downstream storage reservoir. During this operating period the canal system is primarily supply-oriented. The basic objective changes from demand- to supply-orientation; therefore, the applicable operation concept would change from downstream to upstream.

Another example that changes the canal system objectives occurs when a demand-oriented system has a limited or fixed quantity of water supply available at the canal headworks. The limited water supply could be caused by drought conditions or simply an insufficient supply to meet required water demands. The delivery system becomes supply-oriented and the limited water supply determines the canal flow schedule; therefore, the canal must be operated using the upstream operation concept.

1-10. Basic Operational Requirements

The basic operational requirements for a canal system are:

- Deliver water to canalside turnouts
- Transfer specified quantities of water from the upstream source to downstream diversion points at the required times
- Match the inlet supply to the demand for maintaining a balanced canal operation
- Operate within the constraints imposed on the canal system

All canal systems have operational constraints that must be considered when water is being conveyed downstream. Some constraints are:

- Maintain water levels within specified maximum and minimum limits to prevent damage above the canal lining and to provide predictable service to canalside turnouts
- Limit the decrease of water level or drawdown in the canal prism to a specified maximum rate per hour and per day to prevent damage to the canal lining
- Optimize power consumption and pump unit starts per day at relift pumping plants to reduce power and maintenance costs
- Accommodate maintenance schedules

The primary responsibility of operating personnel, when regulating a delivery canal system serving many distribution systems, is to match canal flow to the canalside turnout demand or orders. Scheduling water demand in advance (considering time and quantity) is important to operations personnel. The quantity of water ordered is released from the canal headworks at a predetermined lead time before actual time of delivery. Then the arrival at the downstream turnout coincides with the scheduled delivery time of the diversion. Water users must adhere to their announced schedules and maintain uniform diversions from the distribution system if the canal system is to remain balanced [4].

A canal system that receives its water supply by gravity releases can be regulated closely to maintain a constant flow at the headworks. A canal system that receives its water supply through a relift pumping plant can be regulated only in increments of the smallest pump unit discharge capacity. Water supply regulation is complicated further by the desirability of operating the pumping plant when lower cost offpeak power is in effect, insofar as possible. Under these circumstances, maintaining constant flow in the canal system is virtually impossible. The cumulative daily water orders for the canalside turnouts seldom match the discharge capacity of an integer number of pump units. (Oftentimes, a pump unit needs to be operated for only a portion of the 24-hour period.) Therefore, the pumping plant is either supplying too much or too little water to the canal—depending on the operating period. However, during a 24-hour period the average supply equals the demand.

The actual arrival time of a flow change often differs from the scheduled time at a particular point of delivery. It is difficult for water users to maintain a constant diversion and adhere precisely to their scheduled water orders. The amount of storage in the canal varies with magnitude of flow change situations such as:

- Minor mismatches
- Differences in arrival time
- Variances in turnout delivery
- Storage requirements

All contribute to unsteady flow conditions in the canal [4].

1-11. Supply and Demand Variation

Water scheduling is influenced by several factors that can cause variances in turnout delivery rate. Some factors are automatic pipe distribution systems, on-farm operated valves, sprinkler irrigation, municipal and industrial water service, and water applications based upon on-farm irrigation schedules.

Pumping units located at canalside turnouts, for pressure pipeline distribution systems, often are automated to reduce operating and maintenance costs and to maintain optimum pressure head in the system. Pumping units are started and stopped using pressure, flow, or water level sensors that respond to actual diversions being made by water users.

Usually, ditchriders adjust on-farm turnout valves to maintain diversions in accordance with a schedule that is submitted in advance by the water user. In some instances, this responsibility is being relinquished to the water user to reduce operating costs. Usually, this is a satisfactory arrangement. However, it is reasonable to expect the farmer to make last minute alterations in irrigation scheduling to improve farm operations.

Sprinkler irrigation systems are steadily increasing in use. Sprinkler irrigation increases crop yields and improves efficiency. Sprinkler systems are readily adaptable to simple automatic switching devices that can be programmed to transfer water from one area of land to another in a timely manner. A fully automatic system could be sensitive enough to respond immediately to changes in weather.

Increasingly, water is being delivered through irrigation canal systems to municipal and industrial (M&I) customers. M&I service requires that water be available on demand. Like irrigation systems, many M&I distribution systems have minimal regulatory storage reservoirs.

New techniques such as on-farm irrigation scheduling are available to predict the time and quantity of water needed for the most effective irrigation. Specific amounts of water required for crop evapotranspiration and irrigation, at particular application times, can be derived using the soil-water-plant relationship. Soil characteristics such as infiltration rate and water holding capacity are used in the calculations. By using soil characteristics, moisture content, and estimated evapotranspiration, timing and quantity of water needed to replenish the depleted plant available soil moisture can be calculated and used to predict the next irrigation. Optimum scheduling of water application can be accomplished. Increased irrigation efficiency is obtained through timely water application. Some advantages of using irrigation scheduling are:

- The irrigator knows the specific quantities of water needed and when it should be applied to the crop.
- Human intuition can be eliminated from the irrigation scheduling process.
- The efficiency of the canal system can increase by knowing when specific flow quantities should arrive at the point of diversion.
- The actual timing and/or quantity of water application can be adjusted by considering economic factors involved in the farm operation.
- The actual timing and/or quantity of water application also can be changed at the last moment to account for sudden changes in weather such as rainfall. Other major environmental changes (e.g., solar radiation) may require an earlier or later time of application.

1-12. Automation Applications

Recognizing the benefits of canal system automation is essential in the development and implementation of new principles of water distribution, application, and scheduling. Modern irrigation practices strain the capabilities of the conventional method of canal operation. Canal systems need to respond quickly with greater flexibility to achieve optimum water transfer [5]. Automation can improve efficiency, responsiveness, and flexibility of a canal system. The following examples illustrate possible applications of automation.

In many canal systems, typical carrying capacities of the lower canal pools are only 10 percent of the first canal pool or the design capacity of the canal headworks. A flow variation of 5 percent of the capacity in the upper canal pools could result in a

50-percent variation of flow in the lower canal pools. Prompt corrective measures are required to adjust the intermediate canal check gates and flow at the canal headworks to protect the system. A sudden or unforeseen increase of pumping at a pump turnout could cause serious shortages of water to downstream water users.

Automation of turnouts is another possible application. A turnout diversion to an automatic pipe distribution system is essentially a demand system limited only by the capability of the main canal system to respond. The pumping units and the automatic devices are subject to power and equipment failure. A pumping plant could shut down and the entire diversion would be rejected back into the main canal. A sudden turnout rejection could propagate into the lower canal pools, which usually have smaller carrying capacities, and trigger undesirable operation of the wasteways or cause flooding.

Montoring precise control of diversions is difficult when the user turnout is operated by the water user. Changes in flow caused by sensor activation of automatic sprinkler systems or booster pump rejection would be reflected back into the distribution system and then to the canalside turnout without notification to the canal operator.

A canal system may be required to deliver M&I water on demand at a supply rate that varies more than 50 percent of the average daily demand. This type of turnout would benefit greatly from automation.

An entire system can be automated. Modern irrigation schemes require fast response to flow changes in the canal system. It is difficult for a ditchrider to take prompt and corrective action when flow changes occur frequently, or unannounced. Using automated distribution systems and scientific scheduling techniques, the canal system can respond rapidly to significant variations in the daily diversion rate.

It is possible for a conventionally controlled canal to satisfy operational requirements through application of better methods to predict water schedules. However, precise scheduling is rather difficult to implement on a system basis; it is impossible to predict or anticipate some of the causes of turnout variation such as power and equipment failures. Naturally, water should never be released from the source into the canal system unless a demand exists. If a demand for water is not present, water should remain in the reservoir for later release or use elsewhere. Water stored in a reservoir has economic value as it is beneficial for recreation, increases the hydraulic head for power generation which decreases the energy required for relift pumping in delivery systems.

The main limitation to a conventional canal system operation is the number of visits per day a ditchrider can achieve at canalside turnout(s). Usually, the ditchrider only has enough time to make one or two trips along the canal system during the work day. This limits the flexibility that farmers have to make changes in their water schedule.

Most canal systems do not provide ditchrider coverage during nighttime hours. Usually a ditchrider is available on a standby basis for emergencies. It is absolutely necessary for the water user to adhere to a constant delivery according to the submitted schedule. Any deviation from the schedule likely will affect other water users on the main canal system. However, on some canal systems, the adherence to constant diversion is extremely difficult. In those cases, regular ditchrider nighttime coverage is provided during the peak irrigation season mainly to keep the system balanced and to avoid large mismatches from propagating into the lower canal pools. Normally, water diversions to the canalside turnout are scheduled for a change of flow once daily.

On some canal systems, operations become too difficult to be operated in the conventional manner. Frequent visits to the canal check gate structures are required by the ditchrider to accommodate significant variations in the daily canalside turnout diversion rate. The diversion rate may also deviate significantly from the scheduled daily rate. To provide increased coverage to these turnouts, it probably is necessary to reduce the ditchrider's canal length coverage (circuit). Also, more ditchriders are required on a 24-hour basis to be available immediately to keep the canal system balanced and to avoid large mismatches between the supply and demand. Even with increased ditchrider coverage abnormal operations—caused by sudden and unannounced changes in demand—often are unavoidable.

The typical canal operation maintains the water level upstream of the canal check gate structure nearly constant for all steady-state flow conditions. Therefore, each time canal flow changes to a new steady state, the total water volume in the upstream canal pool must change based upon open channel hydraulics (see ch. 2). Water volume change is referred to as the variable storage requirement and must be considered when flow changes are to be made. Frequent flow changes in the canal system, and thus, frequent changes in the variable storage requirement, make the ditchrider's work even more difficult and usually result in a temporary unbalanced canal operation and response time delay.

The combination of turnout diversion variances, pumping operations, and variable storage requirements burdens the ditchrider's ability to respond to demand changes. Oftentimes, canal pool water levels are allowed to vary. Significant variations in canal pool water levels could cause lining or embankment failure with attendant maintenance costs. Diversion to turnouts on one portion of the system could be satisfied at the expense of diversions on another part of the system. Typically, all turnouts are involved and a compromise is made to satisfy the minimum requirements.

A skilled ditchrider ensures dependable delivery by keeping a surplus flow in the canal. To achieve a surplus flow condition, water levels are allowed to encroach into the canal freeboard. Occasionally, it may be necessary to divert the excess water through a wasteway into a natural cross drainage channel. The excess also could be diverted to a water user who has scheduled an increased diversion at a future time but can accommodate an earlier time of delivery.

CANAL CONTROL HISTORY

1-13. General

The Reclamation Act of 1902 gave authority by the Government to develop water resources for the purpose of supplying irrigation water to the arid 17 Western United States. The Bureau has accomplished an orderly development of water resources in the Western United States resulting in a tremendous increase of agricultural production capacity.

Typically, water sources are located far from the areas where irrigable lands are located. Storage reservoirs have been constructed, which are capable of storing sufficiently large quantities of water, to impound winter surplus runoff to meet annual summer requirements for irrigation and to provide flows for shortages experienced during dry years. Long water conveyance facilities were built to transfer water great distances from the source of supply to places of need. Natural river channels were used when practical to transfer water downstream. Diversion dams have been constructed to elevate and divert water into canal systems. Additionally, water projects provide flood control, power generation, dependable water supply for navigation, salinity repulsion, municipal and industrial uses, fish and wildlife enhancement, recreational opportunities, and other benefits.

The water conveyance system is a major component of the overall water project. The Bureau has constructed about 7,500 miles (12 000 km) of canals through 1986. The flow capacity ranges from 8 to 16,000 cubic feet per second (0.2 to 450 m^3/s). Usually, the major features such as dams are operated and maintained by the Bureau and the distribution systems are maintained by the water users organizations or public irrigation districts.

1-14. Contractual Arrangements

The laws under which the Bureau functions provide the means for owners of irrigable land to obtain financial and technical assistance in the development of irrigation projects. The landowners within the service area form an irrigation, reclamation, or water district under the laws of the state. The Bureau works closely with landowners or water users and the district's elected board of directors to develop the water supply and irrigation facilities to serve irrigable lands. Reports are prepared describing the service area, the quantity of water supply required and its source, the plan of development, and the financial arrangements. These reports are reviewed by the local, State, and Federal levels. After approval by the Secretary of the Interior, the final plan showing engineering and economic feasibility must be presented to the Congress for authorization.

The board of directors elected by the water users of the district has power to tax the land owners and enter into a contract with the Bureau to construct, operate, and maintain the irrigation facilities. Before construction can begin, the district must sign a contract agreeing to repay an appropriate share of the project cost, and the Congress must appropriate necessary funds. The board of directors of the irrigation district represents the landowners throughout these negotiations and transactions. However, before the contract can be executed, a favorable vote of the landowners is required. Additionally, the operating entity must sign an operation and maintenance contract that stipulates how the irrigation system is to be operated and maintained. Typically, this contract provides operating criteria, stipulates when O&M and construction charges will be paid, and when water is to be made available. The Bureau retains title or rights-of-way to all Bureau built facilities.

Construction costs for the repayment are amortized over the repayment period for the project and an annual operating and maintenance cost component is then added to determine the annual cost per acre-foot of water. The water users pay, through the irrigation district or the county tax accessor, for construction, operation, and maintenance costs. The economic feasibility of providing automation depends on the other projects costs. Incorporation of automation in the control scheme development is an important cost and design item.

Water users are entitled to receive a contracted amount of water each year, except during drought years when the quantity may be reduced. Historically, this has been accomplished using scheduled deliveries. Typically, users' submit water orders for their needs 24 to 48 hours before ditchrider(s) make their circuit. Ditchriders accumulate water orders for their circuits and relay these data to district headquarters. The water district accumulates water orders from all circuits and the watermaster establishes a schedule based upon demand and availability. Water users receive their requested amount of water at the time scheduled—providing the accumulated amount does not exceed the capacity of the distribution systems and the supply.

Distribution laterals are not designed to provide service to all water users at the same time because it would be too costly and is not actually required. When daily demands exceed the distribution lateral capacity, the watermaster proportions available water to the water users. Sometimes, individual water users require more water to obtain an efficient irrigation. The water user can move water around the farm or make arrangements with neighbors to irrigate their lands on a rotational basis. The rotation schedule is established working with the ditchrider

The watermaster checks that accumulated orders have not exceeded contracted quantities by turnout and for the system. Oftentimes, weekly and monthly schedules are predetermined by contractual commitments and court decrees. If water orders are greater than established advance schedules, contracted amounts, or the canal capacity, the irrigation district may have to reduce their deliveries and place a restriction on water orders. Then it becomes the responsibility of operating personnel regulating a canal system to match the downstream canalside demands or orders to the upstream available water supply at the canal headworks [6].

1-15 Local Automatic Control History

An early application of local automatic upstream control was that on the Friant-Kern Canal, Central Valley Project, California. The canal is a main conveyance system and delivers a supplemental water supply southerly a distance of 152 miles (245 km), terminating near Bakersfield, California. Originally, it was designed for operation in the conventional manner. However, soon after operations began, the problem of maintaining a constant water level above the White River check gate structure developed.

The canalside turnouts were designed to float on the main canal; i.e., the water level in the main canal established the delivery rate. Any water level variance in the main canal would cause a change in the delivery rate in the distribution system and adversely affect the water users delivery rate schedule. Operating personnel were required to be stationed continually at the check structure to maintain proper service to a nearby major canalside turnout. This was a costly and inefficient operation. Even under continuous vigilance, by operating personnel, undesirable variations in the water level could not be eliminated.

Bureau operating staff realized that an automatic device with sensitivity to water level changes was needed to improve efficiency. They developed a simple automatic device to adjust a check gate whenever the upstream water level deviated more than a predetermined amount using preselected deadband limits from a desired target or set point as shown on figure 1-6. It was named *Little-Man* when the local canal superintendent reported that this successful device was the best "little man" on their staff. The first Little-Man device was put into trial operation in 1952 and was an immediate success. By use of this device, Friant-Kern Canal operation was considered to be automatic for daily flow changes of less than 100 cubic feet per second

Figure 1-6. — Little-Man controller.

(3 m^3/s). Flow changes greater than 100 cubic feet per second had to be conveyed through the canal check gate structures by the ditchriders in the conventional manner.

Sixteen similar devices were installed on other Friant-Kern Canal check gates, and on canalside turnouts. In the latter case, Little-Man controlled the turnout gate from the downstream water level of a Parshall flume used to measure water delivery. The success of the Friant-Kern Canal Little-Man device led to its implementation on many other canal systems. Since the first trial operation—and as experience accumulated—the original design has been modified and improved.

The Columbia River Basin Project (near Ephrata, Washington) implemented the Little-Man on several canal systems for upstream control of the canal check gate structures. However, operations personnel modified the design to achieve a different arrangement of the electrical/mechanical components. The new design included an antihunt feature to prevent control instability and thereby increase the Little-Man's range of automatic control.

The next local automatic controller developed was the device known as the Colvin controller.[2] The controller was a simple electromechanical arrangement of components as shown on figure 1-7. The Colvin controller operates from the water level change in a manner different than the Little-Man. Rather than the Little-Man's preselected deadband limits, a discrete increment of water level change causes a step change in the gate position whenever the discrete increment occurs. The gate step movements are more frequent when the water level is changing rapidly and less frequent if the water level is changing slowly. Therefore, gate movement is a function of the rate of water level change. The variable rate feature can accommodate a wider range of flow disturbances than the Little-Man. Essentially, the Colvin controller made the Little-Man device obsolete.

The Colvin controller control action has application at unattended diversion dams. When storm runoff upstream of the diversion dam causes the reservoir elevation to rise above a set level, the Colvin controller automatically opens the outlet gates to divert surplus flow. Thus, storm runoff is diverted to a storage reservoir downstream before the water level rises above the uncontrolled spillway crest. The Colvin local automatic controller has many applications for either upstream or downstream control for

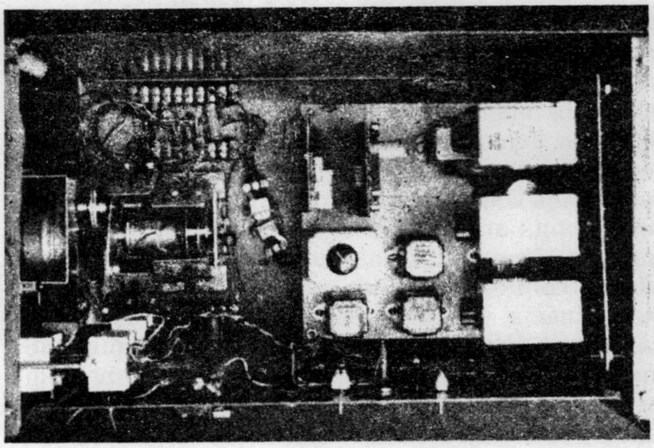

Figure 1-7. — Colvin controller.

the single control structure such as the diversion dam sluice gate or the canalside turnout.

Both the Little-Man and the Colvin controllers have limited applications for automatic regulation of a series of canal check gate structures. Both devices are limited to controlling only the smaller canal system flow changes to avoid instability of the entire canal system [7].

Recognizing the limitations of the Little-Man and the Colvin controllers for canal automation and the need to include available control theory technology, the Bureau's Mid-Pacific Regional office (Sacramento, Calif.) initiated an ongoing research and development program. The program began in 1966 with a research and development contract with the University of California at Berkeley [8, 9].[3]

The initial investigation developed a mathematical model for canal system simulation and an analytical study for the selection of control parameters. From the results of the 1966 research program, which ended in 1969, four prototype analog proportional controllers known as the hydraulic filter level offset (HyFLO) method were designed, constructed, and field tested for automatic downstream control. The controllers incorporated certain control parameters into additional elements of the feedback system designed to avoid instability and thus achieve a higher degree of self regulation. The feedback system was named the hydraulic filter level offset method because the main element necessary to have control stability involved a hydraulic filter well. The hydraulic filter well provides the time lag compensation

---

[2] The controller was developed by William Colvin (retired) of the Bureau of Reclamation in 1971.

[3] The principal investigator was Professor James A. Harder; Micahel J. Shand, doctroral canidate, performed the investigative studies and wrote two final reports—as required by the contract.

necessary for downstream control applications. The HyFLO method was designed to achieve a fast response and recovery of the canal system to a new steady state without excessive overshooting of the water levels [10].

Problems developed with the hydraulic filter well operation. Debris (dust and dead insects) plugged the small capillary tube entrance to the filter well from the inside. As a result, it was decided to develop a reliable electronic time delay circuit to replace the more cumbersome and costly hydraulic filter well. A float-tape-pulley arrangement still was used to operate a potentiometer for measuring canal water level changes. The controller employing the electronic filter was renamed to electronic filter level offset (EL-FLO). The electronic filter successfully replaced the hydraulic filter element of the feedback system.

Prototype tests on the Corning Canal provided the basis for developing a new prototype EL-FLO controller and electronic filter unit suitable for permanent installation. The new design incorporated the latest technology in solid-state circuitry and construction. The objective was to achieve better equipment reliability. Also, it was decided to develop a RESET controller and incorporate it into the new prototype EL-FLO controller. The RESET feature would eliminate the residual water level offset characteristic of the proportional mode of the EL-FLO method.

The new prototype EL-FLO plus RESET analog controller chassis, as shown on figure 1-8, was thoroughly tested in the laboratory and in the field on the South Gila Canal. The equipment testing highlighted two important concepts:

- An alarm system must be an integral part of the control system to notify operating personnel at the watermaster's headquarters immediately when abnormal operating conditions developed because of equipment failure or malfunctions.
- Fail-safe circuitry must be included as much as possible to prevent further gate movement when equipment fails such as when the input signal is lost.

The results of the prototype development program of the EL-FLO plus RESET controller—including laboratory and field testing [11]—served as a basis for designing and writing specifications for control units to be procured and installed on a permanent basis.

The EL-FLO plus RESET automatic downstream control system was installed on the Corning Canal involving 12 check structures, and on the Coalinga

Figure 1-8. — EL-FLO plus RESET controller.

Canal (near Coalinga, Calif.) involving 3 check structures. These controllers were put into operation in 1974. The experience gained from continuous operation of these two canals has pointed out an important factor involved in a successful operation. An experienced technician with an electronic background is indispensable to perform the necessary maintenance on the control chassis, alarm system, and the communication channels.

Recognizing the potential applications to single control structures, further development of the Colvin controller continued from 1977 to 1980. With the original device, reliability problems were experienced with the mechanical components after relatively short periods of operation. A reliable electronic design was developed and tested in the laboratory. The electronic design included a reset feature that would continue gate opening adjustments on the occasions when a steady-state water level was established outside the deadband limits—an inherent characteristic of the Colvin controller.

Field test indicated that greater gate movements in response to the higher rates of water level changes were needed. An improved design incorporated a two-step gate run-time increment, small and large. The larger step-gate run time would be used when the water level changed rapidly. The modified electronic design was field tested on the Loveland turnout from the Hansen Feeder Canal, (near Loveland, Colo.) for downstream control. It was placed into operation during the 1979 irrigation season and has continued to operate satisfactorily.

An improved electromechanical device was developed during the latter phase of the Colvin controller laboratory development and testing program. Although this design was more expensive (it required custom made parts), it was simpler than to operate than the electronic design and was easier to maintain. It was field tested as an upstream controller on the North Poudre supply canal diversion dam sluice gate (near LaPorte, Colo.) in 1980. The device, shown on figure 1-9, continues to operate successfully.

Further development of local automatic controllers involved the application of the digital microprocessor based equipment which became available at a reasonable cost in the 1980s. The microprocessor can be programmed with the necessary control algorithm and replace the more cumbersome analog control equipment. The first application of a prototype microprocessor programmed with the Colvin algorithm for local automatic downstream control was made at the Flatiron afterbay outlet gate (near Loveland, Colo.). The microprocessor equipment operated satisfactorily for an irrigation season.

Figure 1-9. — Improved Colvin controller.

However, field operating personnel opted for the simpler electronic/mechanical device which they could easily maintain. Additional technical expertise would have been required to maintain the microprocessor equipment and could not be provided within an acceptable benefit/cost ratio.

As of 1986, Colvin controllers are being used on three diversion dams of the San Juan-Chama Project located in southern Colorado and northern New Mexico. A program has been implemented to replace the EL-FLO plus RESET analog control chassis beginning with the first four check structures on the Corning Canal [12]. Digital equipment has made most analog equipment obsolete.

1-16. Supervisory Control History

The first major effort to implement a supervisory manual control system was on the Delta-Mendota Canal (near Tracy, Calif.). It was originally designed to be operated in the conventional manner. In 1968, a supervisory manual control system (using solid-state components for the first time) was placed into operation to effectively meet new operating requirements imposed on the canal system by the addition of the Bureau's San Luis Unit. The San Luis Unit is an offline reservoir that stores off-season surplus flow which occurs in the San Joaquin-Sacramento River Delta at the Delta-Mendota Canal headworks. Off-season capacity of Delta-Mendota Canal is used to transfer surplus flow into the San Luis Unit.

Inflow to the Delta-Mendota Canal has six large centrifugal pumping units. Therefore, input to the Delta-Mendota Canal can only be regulated in increments of one pump unit discharge. The problem of input regulation is further complicated by the desirability to operate with offpeak power, insofar as possible. An additional operational requirement exists when transferring large quantities of water during the off-season into the San Luis Unit. To accomplish the upgraded operation, the canal system required supervisory control from a centralized location. Supervisory manual control type of operation system was selected and installed (in 1968) on the Delta-Mendota Canal for the first 70 miles (113 km), which included 13 canal check gate structures and the O'Neill Forebay Pumping Plant. The Delta-Mendota Canal supervisory manual control system equipment has been updated with new technology equipment.

The second major supervisory manual control system placed into operation was on the Navajo Canal, Navajo Indian Irrigation Project, (near Farmington, New Mex.). An initial investigation [13] determined, through mathematical model simulation studies, that local automatic control would not provide satisfactory control of the canal system. Studies demonstrated that local manual control would be difficult primarily because certain sections of the canal system did not have enough canal check gate structures; i.e., the spacing of the check gate structures was too far apart. The investigation recommended supervisory manual control as the primary control scheme. Supervisory control system was placed into operation in 1985. The installation is different from the Delta-Mendota Canal System in that a very high frequency (VHF) radio system is used for required two-way communication channel between the headquarters and the canal check gate structures.

The most comprehensive application of supervisory control for the operation of a canal system to date has been the control system for the CAP (Central Arizona Project). The CAP is a large aqueduct system which delivers Colorado River water to urban, agricultural, and industrial water users in central and southern Arizona. Up to 3,000 cubic feet per second (85 m^3/s) is conveyed 190 miles (300 km) from Lake Havasu on the Colorado River. The project has 14 inline pumping plants with over 100 pump units, 37 dual gate check structures, and over 40 turnout structures.

The basic operational supervisory control objectives of the CAP are stability, responsiveness, and favorable economics [14]. These objectives are all applicable from both the operator's and the water user's concern. The major operational constraints on the system are:

- Aqueduct size and length
- Number and location of pumping plants and check structures
- Operating limits on water level variance
- Lack of regulating reservoirs or wasteways

Little margin for abnormal operations exists in the system and the appropriate control of flow changes is a significant factor.

To satisfy the operational objectives within the limitations imposed by the constraints, a control volume method of operation was chosen [15]. The control volume operation allows rapid response to user demand with less water level fluctuation than other methods of operation. Major economic savings could be realized by minimizing the number of pump units starts. These operations require sophisticated scheduling software capable of manipulating large quantities of data and making complicated control decisions.

The aqueduct system uses both supervisory automatic and computer directed control methods. Operation of the entire aqueduct system is controlled by a programmable master supervisory control (PMSC) system that consists of a dual computer master station, intelligent RTUs at the field sites, and a redundant communication system. The PMSC operates as a computer-directed supervisory control system that automatically develops control actions in the master station and sends them to the RTUs at the field sites. The RTUs have the capability to provide automatic control at the check gate and pumping plant locations.

The Bureau's Arizona Projects Office has combined innovative ideas with the latest technology in the development of the supervisory control system; an automated operation that is highly sophisticated has resulted.

1-17. Summary

The Bureau works closely with the water user and water districts during design, construction, and operation of the main canal and distribution system. The Bureau has been concerned with the efficient transfer of water since the first canal was placed in operation to meet the contractual agreements and daily water schedules. Field operation staff recognized the need for canal automation. The first major effort was the development and implementation of the Little-Man device for local automatic control of the Friant-Kern Canal check gates in 1952. Its immediate success, even though limited to the regulation of small canal flow changes, led to application on other canal systems and the development of newer technology.

In 1971, the Bureau's Denver Office formed a water systems automation team. The team consists of various engineering and scientific disciplines and includes a working group. The team and the working group were given the responsibility to assist in the ongoing canal automation program. Since then, an extensive research, development, and implementation effort has been progressing both in the Denver Office and in the Bureau's regional offices.

It is evident from the history of local automatic control that extensive laboratory and field prototype testing is involved whenever a new control concept or new equipment is applied to canal system control. A canal system is difficult to regulate and the canal bank location imposes many environmental hazards on the equipment. The theoretical control concept must be correctly applied to the actual canal operation if the goal of optimum transfer of water is to be achieved. Control equipment hardware must withstand the canal bank environment and function without failure for long periods of continuous operation.

The Bureau's efforts in the ongoing canal automation program has completed much of the necessary research and development—including laboratory and field testing—for the control systems just described. Advances in equipment technology are currently being developed. Undoubtedly, field experience is required with new hardware to eliminate potential equipment problems. The available and successful control techniques discussed have been proved by actual operations and do not require further major verification test programs. Advances in technology likely will not cause these control concepts to become obsolete.

Limitations of operating when using the conventional manner have stimulated the Bureau to adapt local automatic and supervisory control methods of operation to main canal systems. The program is directed toward modernizing and upgrading the conventional methods of operation. The objective is to capitalize on the benefits realized from local automatic and supervisory control canal systems operations. The benefits would primarily be better service to the water user, efficient transfer of water from the source to the point of use, and reduced canal system operating cost. The experience gained in this ongoing program should influence the design of future projects. By recognizing the advantages and the flexibility for applying modern control concepts to canal system operation (in the early planning stages) additional significant benefits could result to new projects.

BIBLIOGRAPHY

1-18. Bibliography

[1] Chow, Ven Te; *Open-Channel Hydraulics*, McGraw-Hill Book Co., New York, 1959.

[2] Aisenbrey, A. J., Jr., R. B. Hayes, H. J. Wamen, D. L. Winsett, and R. B. Young, *Design of Small Canal Structures*, Bureau of Reclamation, Denver, Colorado, reprint 1987.

[3] Buyalski, Clark P., "Discharge Algorithms for Canal Radial Gates," Bureau of Reclamation Report No. REC-ERC-83-9, 232 pp., Denver, Colorado, December 1983.

[4] Sullivan, Edwin F., and Clark P. Buyalski, "Operation of Central Valley Project Canals," a paper presented at the ASCE Irrigation and Drainage Division Conference, Phoenix, Arizona, November 13-15, 1968.

[5] Harder, James A., Michael J. Shand, and Clark P. Buyalski, , "Automatic Downstream Control of Canal Check Gates by the Hydraulic Filter Level Offset (HyFLO) Method," a paper presented at the Fifth Technical Conference, U.S. Committee on Irrigation, Drainage and Flood Control (U.S.I.C.I.D.), Denver, Colorado, October 8-9, 1970.

[6] Buyalski, Clark P., "Automatic Downstream Control of the Corning Canal Check Gates by the EL-FLO Plus RESET Control Systems," a paper presented at the Symposium for U.S./U.S.S.R. Cooperative Program, *Automation and Remote Control of Water Resource Systems*, Frunze, Kirgiz, S.S.R., May 1979.

[7] Buyalski, Clark P., "Basic Equipment in Automatic Delivery Systems," a paper presented at the National Irrigation Symposium, *Irrigation Today and Tomorrow*, Lincoln, Nebraska, November 10-13, 1970.

[8] Shand, M. J., "Final Report—The Hydraulic Filter Level Offset Method for the Feedback Control of Canal Checks," Technical Report No. HEL-8-3, *Computer Program for System Simulation*, University of California, Hydraulic Engineering Laboratory, Berkeley, California, June 1968.

[9] Shand, M. J., "Final Report—Automatic Downstream Control Systems for Irrigation Canals," Technical Report No. HEL-8-4, *Selection of Control Parameters for the Hydraulic Filter Level Offset (HyFLO) Method*, University of California, Hydraulic Engineering Laboratory, Berkeley, California, August 1971.

[10] Schuster, J. C., and E. A. Serfozo, "Study of Hydraulic Filter Level Offset (HyFLO) Equipment for Automatic Downstream Control of Canals," Bureau of Reclamation Report No. REC-ERC-72-3, 16 pp., Denver, Colorado, January 1972.

[11] Buyalski, Clark P., and Edward A. Serfozo, "Electronic Filter Level Offset (EL-FLO) Plus RESET Equipment for Automatic Downstream Control of Canals," Bureau of Reclamation Report No. REC-ERC-79-3, 149 pp., Denver, Colorado, June 1979.

[12] Buyalski, Clark P., "Navajo Indian Irrigation Project (NIIP)—Evaluation of Control Schemes," attachment to internal memorandum to Water Systems Automation Team Leader, code D-430 from Chief, Division of Research and Laboratory Services, Bureau of Reclamation, Denver, Colorado, January 10, 1979.

[13] Falvey, H. T., and P. C. Luning, "Gate Stroking," Bureau of Reclamation Report No. REC-ERC-79-7, 145 pp., Denver, Colorado, July 1979.

[14] Graves, Albert L., and Robert S. Gooch, "Central Arizona Project Startup," a paper presented at the ASCE, Water Forum '86: World Water Issues in Evolution, Long Beach, August 4-6, 1986.

[15] Mattaei, Alfred, "Corning Canal and Red Bluff Diversion Dam Microprocessor Controller Report," Bureau of Reclamation, Willows, California, March 1987.

CHAPTER 2

OPERATIONS

INTRODUCTION TO CANAL OPERATIONS

2-1. Principles of Operation

a. **Definition**.—The canal system operation transfers water from its source(s) to one or more points of diversion downstream. Operations deal with the movement and behavior of water in a canal system, and are dependent on the principles of open channel hydraulics. The primary function of operation is to manage the changes in flow and depth throughout the canal system. The term "operation" refers to the hydraulic reaction in the canal pools which results from control actions.

b. **Objectives**.—Basically, canal systems are built for the purpose of conveying water from one place to another. The most prevalent reason for water transfer is to irrigate farmlands, with the objective of helping farmers increase crop yields. Canals also convey water for

- M&I (municipal and industrial) use
- Storm and snowmelt runoff to natural drainage channels
- Collecting water from several independent sources into a single supply
- Electrical power generation
- Fish and wildlife
- Recreation

The objective of building and operating a canal system is to serve the above purposes as efficiently and economically as possible. To do this, canal operations should be tailored for each canal system to fulfill that system's specific requirements. But the requirements imposed on canal systems are steadily becoming more complex. Modern irrigation practices require greater delivery flexibility and increased efficiency. The current higher costs of water, power, and construction decrease the feasibility of wasting water from canals or building oversized canals to obtain additional in-channel storage. The use of intermediate storage and wasteways is being curtailed by environmental assessment. Because of these and other factors, canal systems need to respond quickly to flow changes. The overall canal operation must be improved to provide this capability.

Canal system automation should not be thought of as an end, but rather as a means to accomplish the desired canal system operations. The true goal should be to achieve the most efficient and beneficial operation possible. Expanding control system capabilities is one method to economically reach this goal.

c. **Balanced pool operation**.—As defined in chapter 1, the length of canal between check structures is called a canal pool. Occasionally, an inline pumping plant may serve as a pool boundary instead of a check structure. When total flow out of the canal pool equals total inflow, the pool operation is balanced. Referring to figure 2-1, a balanced pool operation exists when:

$$QC_1 = QTO_1 + QC_2$$

where:

QC_1 = pool inflow
QTO_1 = turnout deliveries from the pool
QC_2 = pool outflow at the downstream end

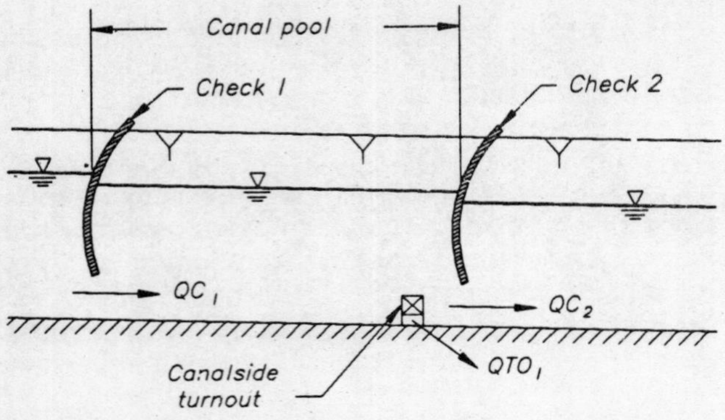

Figure 2-1. — Balanced pool operation.

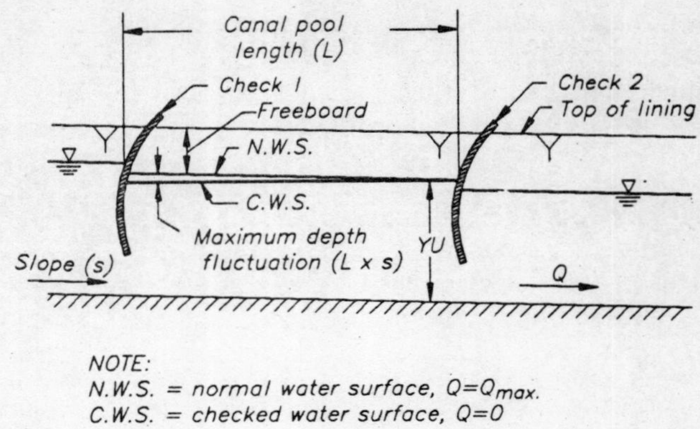

Figure 2-2. — Conventional canal operation.

d. Conventional operation.—Traditionally, canals have been sized to convey the maximum design flow with freeboard added as a safety factor against excess water depths. Canals of this design usually have been operated in the "conventional" manner. Figure 2-2 illustrates some of the characteristics of conventional operations.

An important characteristic of conventional operation is the attempt to maintain a constant water depth at the upstream side of each check structure. When flow changes, the water surface profile within each canal pool essentially pivots about this constant depth at the downstream end of the pool. Conventional operation has had such widespread use for so many years because (1) construction costs can be minimized by sizing the canal prism only large enough to convey the maximum steady flow, and (2) the canal system operation can be successfully accomplished using local manual control.

e. Supply and demand.—Canal operation concepts (see sec. 1-9) are based upon whether the water schedule priorities are supply- or demand-oriented. Supply refers to the quantity of water that is available or being provided from an upstream source. The upstream operation concept applies to canal systems that are primarily supply-oriented. A supply-oriented system, therefore, must be operated to satisfy upstream requirements. One example of this is a storm drainage system.

Demand refers water needed or being delivered at points downstream in the canal system. The downstream operation concept applies to canal systems that are primarily demand-oriented. These systems must respond to downstream conditions, and the water supplied from upstream is a function of downstream demand. A municipal water system is an example of a demand-oriented system. Water users at downstream ends of the system take water without restriction; the operation responds only to downstream conditions.

f. Prediction versus reaction.—Some canal systems only react to circumstances after they happen, while others are operated by anticipating situations in advance by predicting the future demands on the system. Automatic controls are purely reactive as they respond to present or past information. However, prediction often is used in manually controlled systems to enhance the operation; it can be used in automated systems which include manual input.

The more common application of canal operation using prediction is in scheduled delivery. Operations scheduling is based upon water orders being turned in by water users prior to the actual delivery. Anticipation also can be based upon historical data, daily or weekly trends, meteorological information, or agricultural data such as soil moisture content.

2-2. Canal Systems

As discussed (sec. 1-5), a canal system can be classified according to its principal objectives as a delivery system, collector system, or connector system. Certain characteristics of each of these types of canal systems will affect operations. In addition to the different priorities associated with supply and demand, physical differences exist which need to be considered.

A delivery system has a large capacity at the upstream end and gradually decreases to a small capacity at the downstream end—sometimes branching into successively smaller lateral canals. Water is turned out at various points along the length of the canal as it tapers to the final turnouts at the extreme end(s). This decrease in capacity amplifies the effects of a flow mismatch as it moves downstream.

For example, consider a canal having a capacity of 500 cubic feet per second (14 m^3/s) at the upstream end and tapers to a capacity of 100 cubic feet per second (2.8 m^3/s) in the downstream pool. If a mismatch of 25 cubic feet per second (0.7 m^3/s) occurs at the upstream end, this is only a 5 percent error. This amount of error may be normal and acceptable without creating significant problems in the upper pools. But as the 25 cubic feet per second mismatch is passed downstream it will become significant in the downstream pools where it will be a 25 percent flow error.

Water users at the downstream end of delivery systems, referred to as "tail-enders," often will suffer from too much or too little water. Normally, surplus water must be supplied at the headworks to prevent water shortages to tail-enders. Oftentimes, surplus is wasted at the downstream end of the system. Typical waste in a conventionally operated canal system is about 5 to 10 percent of the total inflow [1].

This problem stems from combining the downstream operation concept (demand-oriented) with the upstream control concept. As noted in chapter 1, this is an inefficient combination. When using the upstream control concept, a large portion of the flow discrepancies tend to be passed downstream. These problems end up at the system's tail-end, where more difficulty is encountered dealing with them.

Delivery systems will operate more efficiently with a downstream control concept. Flow mismatches will be passed in the upstream direction into successively larger pools. The effects of a flow mismatch will be attenuated as it moves upstream.

The same reasoning can be applied to the exact opposite situations for collector systems. Since collector systems tend to increase in capacity in the downstream direction, operations will be most efficient when the upstream control concept is applied.

Connector systems will be neutral to these concepts because flow capacity should remain constant throughout the canal length. For a connector system, the source of the major flow changes should determine which control concept is to be used. If flow changes generally originate at the upstream end, then the upstream control concept will work best. The downstream control concept should be used if most flow changes originate at the downstream end.

The preceding discussion is particularly important when automatic control is involved. These considerations will help to identify the types of automatic control systems that best suit a particular canal. Generally, a canal system's operations must be defined clearly before automatic control can be successfully applied. This will make some canals better candidates for automation than others.

Physical characteristics also affect the potential for successful automation. An example is a steep channel slope. A canal having a steep bottom slope and/or many vertical drops in the invert will be more difficult to automate than one having gentle slopes. If supercritical flow exists, the upstream flow is no longer influenced by downstream conditions. This will prevent typical depth-based automatic controllers from functioning effectively.

The storage capacity within the canal prism is another important physical characteristic that influences canal operation and automation. Typically, flow changes are accompanied by changes in water volume stored in the canal because there will be different water surface profiles at different flow rates. These principles will be further developed later in this chapter.

2-3. Water Schedule

The water schedule for a canal system is the flow versus time relationship. This can range from nearly constant flow—for a canal with little flow change—to a complex schedule of flow rates that change frequently at numerous points throughout a large canal system. Every canal has a water schedule, but a variety of ways to predict, create, and implement the schedule exist [3, 4]. Usually, the water schedule is predicted or arranged in advance and then adjusted during implementation. The flexibility to adjust the water schedule can vary widely from one canal to the next.

Schedules often are addressed in conjunction with delivery. Delivery refers to release of water from the canal system to the water user. This release usually occurs at some type of turnout structure. A delivery schedule is the quantity of water to be transferred to a user, as a function of time. Scheduled delivery is one type of delivery concept. Delivery concepts are discussed later in this chapter.

A conventionally operated canal is used as an example to explain the water schedule. With the conventional method, the water schedule specifies:

- Quantity of water to be transferred downstream
- Location where specific quantities are to be diverted
- Time of the diversion

The development of the water schedule typically begins at least one day before the actual demand requirement. Individual users submit water orders to their ditchrider or to their district headquarters. The water district compiles the individual water orders for the distribution system associated with each canalside turnout from the main canal system.

Sometimes the water district will modify individual water orders when it is impractical or impossible to meet all requests. Usually, modifications involve limiting all individual orders by a particular percentage. Another possibility is to change the timing of deliveries to avoid large flow changes in the canals.

After all individual orders are collected, the ditchriders submit new canalside turnout orders to the watermaster of the main canal system. Then, the watermaster accumulates the canalside turnout orders by canal pool. Flow requirements for each canal pool are summed beginning at the most downstream pool as shown on figure 2-3.

Canal flow requirement by canal check would be:

$$QC_3 = QTO_{3.0} + QC_4$$
$$QC_2 = QTO_{2.0} + QTO_{2.1} + QC_3$$
$$QC_1 = QTO_{1.0} + QC_2$$

where QC_4, QC_3, QC_2, and QC_1 are inflows into canal pools 4, 3, 2, and 1, respectively, and $QTO_{3.0}$, $QTO_{2.0}$, $QTO_{2.1}$, and $QTO_{1.0}$ are turnout demands.

The sum of the canal pool flows progresses upstream to the canal headworks, check 1. Therefore, the water schedule for the canal system is established for the following day. The schedule has the total input flow required at the headworks, all the canalside turnout flow diversions, and the flow required at each successive canal check gate downstream to meet the turnout flow changes en route.

Some canal systems operate with water delivery to users *on-demand*; the water schedule is not arranged in advance. In these cases, the schedule often can be predicted in advance based upon typical water use patterns and any other known circumstances which influence water use. The water schedule then will need to be adjusted to a "real-time" environment, in reaction to changing demands.

2-4. Water Transfer Procedure

In a conventional canal system, the ditchrider adjusts the check gates to satisfy the water schedule. First, the ditchrider must determine the new canal check gate openings required for the flow changes. An experienced ditchrider is able to predict the relationship between gate opening and flow with reasonable accuracy. Usually, the ditchrider begins to implement the new water schedule in the early morning by changing the flow at the head of the canal. The headgate opening is adjusted to obtain the new total flow, QC_1. Then, the ditchrider follows the flow change downstream to make the required adjustments at canalside turnouts and the successive check gates—sequentially.

The check gates usually are adjusted to maintain a desired depth on the upstream side of the check structure. This is the upstream control concept. This helps to achieve steady flows through gravity turnouts which are located just upstream from the check structure. As the flow change progresses downstream through each canal pool, the ditchrider will attempt to achieve a balanced pool operation.

Figure 2-4 shows a flow increase progressing downstream through the pool. When the flow change reaches the turnout, the turnout flow will be adjusted to the scheduled value. Then, the remaining portion of the flow change will proceed past the turnout to the next downstream check gate, check $n+1$.

When the flow change arrives at check $n+1$, the check gate opening is adjusted to transfer the flow mismatch (between the upstream canal pool inflow and turnout outflow) into the next downstream canal pool. If check $n+1$ is adjusted properly, pool inflow and outflow will be balanced and the check $n+1$ upstream depth, *YU*, will remain relatively constant. A new steady-state flow is established as shown on figure 2-5.

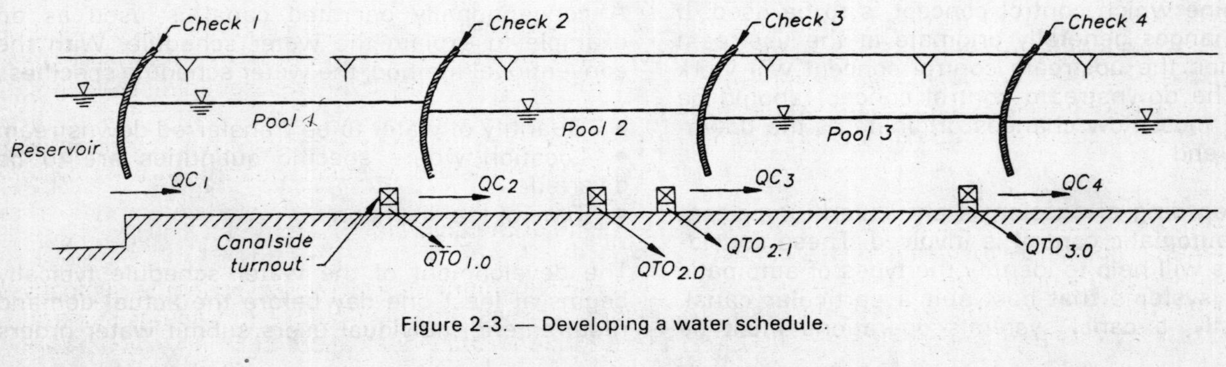

Figure 2-3. — Developing a water schedule.

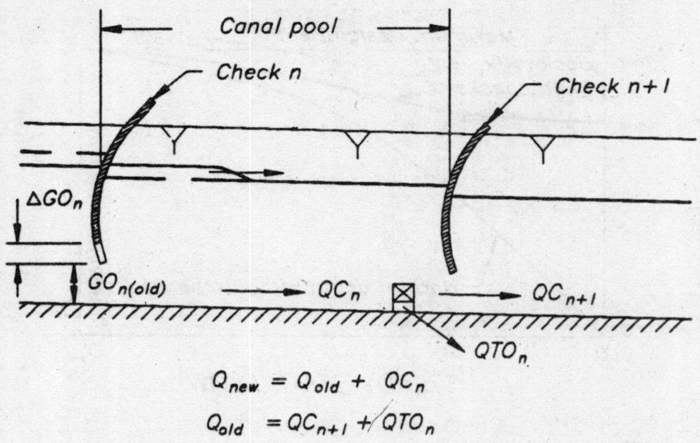

Figure 2-4. — Flow increase progressing downstream.

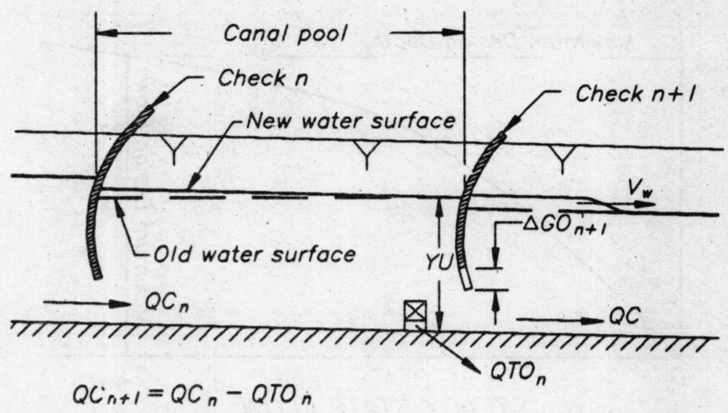

Figure 2-5. — Downstream gate adjustment to transfer flow change.

The sequence for making flow changes and balancing the canal pool operations then continues downstream until all turnout flow changes have been made according to the water schedule. To manually correct flow imbalances with just a single gate adjustment, at each check structure, is difficult. Realistically, regulation is more difficult than it appears in this idealized example. Hence, only one trip down the canal by the ditchrider is often insufficient. Typically, subsequent gate adjustments are required to balance flow conditions.

The above procedure is one example to illustrate some of the details in transferring water through a canal system. It is a typical procedure for many conventionally operated canals. Other water transfer procedures will be used for canal systems that are operated and controlled by different methods.

CANAL HYDRAULICS

2-5. Flow Types

Depending upon how canal flow depth changes with respect to distance and time, the flow can be classified into several different types.

- Changes with respect to *distance* are referred to as *varied flow*.
  - Gradually varied flow, if the depth changes gradually over a long distance
  - Rapidly varied flow, if depth changes are abrupt over a short distance
- Changes in with respect to *time* are referred to as *unsteady flow*.
- If the depth does not change with time, flow is *steady* (steady-state).

All flow types can be summarized as [5, 6, 7]:

Steady Flow
- Uniform flow
- Varied flow
  - Gradually varied
  - Rapidly varied

Unsteady Flow
- Uniform flow
- Varied flow
  - Gradually varied
  - Rapidly varied

These flow types are described in the following sections. The important characteristics of these flow types relative to canal automation are considered separately.

2-6. Steady Flow

a. Steady, uniform flow.—This flow type is characterized by a constant depth in each pool. The condition exists at maximum design flow with all check gates wide open as shown on figure 1-3. Steady uniform flow is also called flow at *normal* depth. With steady uniform flow, the friction force on the sides and bottom of the canal prism is exactly balanced by the component of force due to gravity acting in the direction of flow. The normal depth in a canal depends upon the flow boundary roughness, canal geometry, and flow rate. For a given roughness and geometry, the normal depth in a canal pool can be determined as a function of flow as shown on figure 2-6.

b. Steady, gradually varied flow.—This flow type is characterized by a change in the depth with distance along each pool. Steady, gradually varied flow is an important flow condition in the design of automated canals. One type of steady, gradually varied flow in canals is that formed upstream of check structures as shown on figure 2-7. In this case, the steady, gradually varied flow profile is known as the *backwater profile*. The description of other types of steady, gradually varied flow profiles can be found in [5].

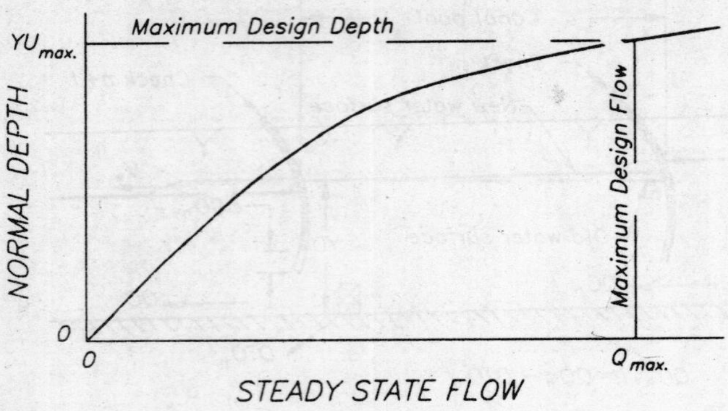

Figure 2-6. — Normal depth relationship in a canal.

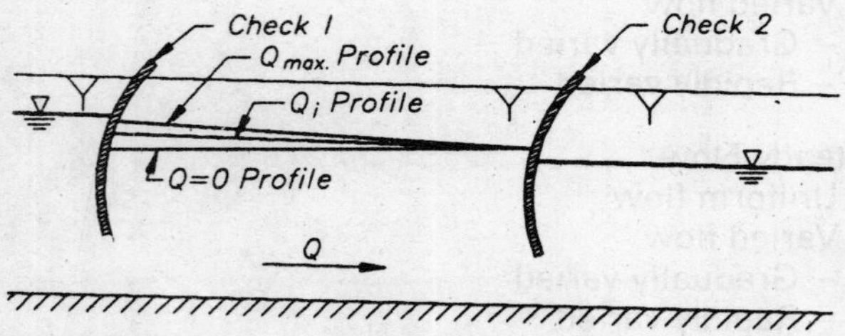

Figure 2-7. — Backwater surface profile.

The capability to operate a canal would be limited severely without the backwater formed by check structures. If check structures were not used to regulate canal flow, water depth in the canal prism would be a function of flow rate, as shown by the normal depth curve on figure 2-6. At low flows, canal water depth could be too low to supply canalside turnouts. Additionally, flow depth would vary greatly as flow rate changed. The backwater surface profile formed by the check gates lies between the full-flow profile and the zero-flow profile, as shown on figure 2-8. Thus, for a given flow, depth in the pool can be substantially greater than the normal depth that would have existed without the backwater.

c. Steady, rapidly varied flow.—This flow type occurs below check gates during free flow operation. The flow immediately downstream of the gate and the flow at the hydraulic jump are both examples of steady, rapidly varied flow. Generally, steady rapidly varied flow should be avoided in automated systems because it introduces a discontinuity in flow which is difficult to include in the control algorithm.

2-7. Unsteady Flow

a. Unsteady, uniform flow.—For this flow type, water level changes with time, while remaining parallel to the canal invert. Because this condition is impossible to achieve in the field, it will not be considered further.

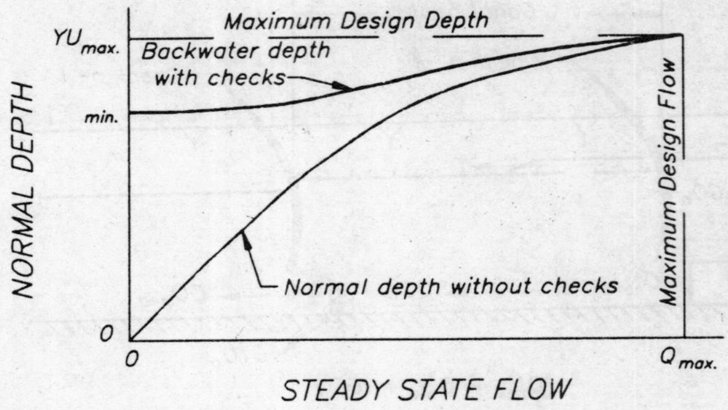

Figure 2-8. — Checked water depth relationship.

b. Unsteady, gradually varied flow.—With respect to operating automated canals, this is an important flow condition. For example, a gate movement produces changes in flow and in adjacent depths as shown on figure 2-9. A flow change that is initiated by changing the check gate opening takes the form of traveling translatory waves. A *translatory wave* is a gravity wave that propagates in an open channel and results in displacement of water particles in a direction parallel to the flow [5].

For example, if a check gate is opened a small increment, as shown on figure 2-9, flow into the downstream canal pool will increase—generating a positive translatory wave which will travel downstream. Simultaneously, a negative translatory wave is generated upstream of the check gate; this wave travels upstream. Translatory waves also are generated when the check gate closes. However, in this case, a negative translatory wave progresses in the downstream direction and a positive wave progresses upstream. Figure 2-10 is a schematic of a translatory wave profile.

The amplitude (height) of the translatory waves is a function of the magnitude of flow change and canal prism geometry. Friction forces will attenuate or decrease the height of the wave front as the wave transverses the entire canal pool. Theoretically, the translatory wave has the following features:

- Successive positions of the wave front at different times are parallel to each other (fig. 2-10).
- Wave velocity, or *celerity*, is greater than the mean canal water velocity.
- Wave configuration travels at a constant velocity, but the mean water velocity may vary from pool to pool.

The translatory wave velocity is:

$$V_w = \frac{Q_u - Q_d}{A_u - A_d}$$

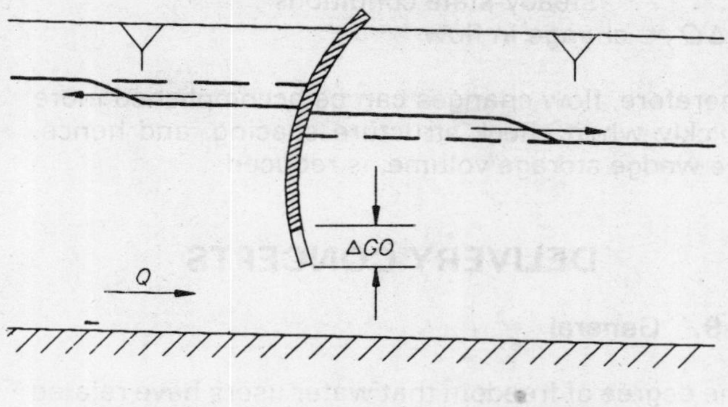

Figure 2-9. — Flow change due to gate movement.

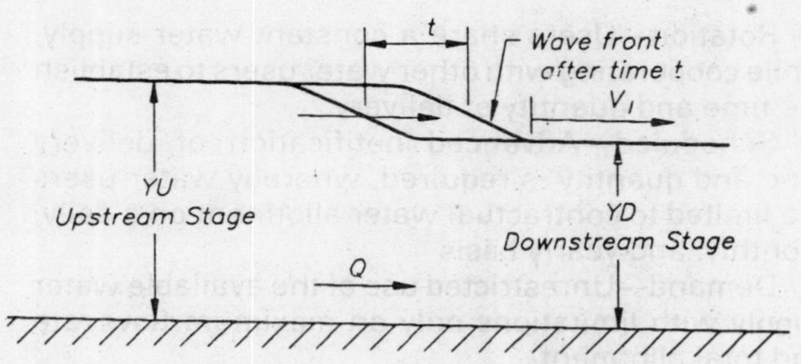

Figure 2-10. — Translatory wave profile.

where:
A_u = upstream flow cross-sectional area
A_d = downstream flow cross-sectional area
Q_u = upstream flow rate
Q_d = downstream flow rate

From a practical viewpoint, the translatory wave changes shape because the wave's leading edge travels at a velocity that is also a function of the canal prism geometry. Velocity of the leading edge of the wave front is:

$$V_1 = V_m \pm C$$

where:
V_1 = velocity of leading edge of wave
V_m = mean flow velocity = Q/A
Q = flow rate
A = canal prism cross-sectional area
C = wave celerity = $(gD)^{1/2}$
g = gravitational constant
D = hydraulic depth = A/T
T = wetted prism top width

A positive sign is associated with waves moving downstream and a negative sign with waves moving upstream. The leading edge of the translatory wave travels much faster than the final wave velocity. Depending upon the value of the hydraulic depth, the wave velocity may be as much as 10 times the mean water velocity.

The velocity of the translatory wave is an important operating criteria; it determines when the effects of a flow change at one point in a canal will reach other points. For instance, a common assumption is that a flow increase (at a canal headworks) will travel downstream at the flow velocity in the canal, and that flow increases at canalside turnouts cannot be accomplished until this additional water arrives. In reality, flow will begin to increase at points downstream as soon as the leading edge of the translatory wave front arrives, which is much sooner than would be predicted from the average flow velocity in the canal.

The water surface behind the translatory wave slowly approaches the steady, gradually varied flow profile. The amount of time required to fully achieve the steady profile is a function of:

- Initial flow
- Final flow
- Channel properties
- Check structure spacing
- Initial water depths
- Final water depths

These factors will be discussed later in this chapter.

c. Unsteady, rapidly varied flow.—If check gate motion is too abrupt or if the front of the translatory wave becomes too steep, a bore wave will be formed. A bore can be thought of as a travelling hydraulic jump. The formation of bores should be avoided in automated canals because they produce large water surface fluctuations at control structures and undesirable disturbances in the control for automated gates. Presently, computer programs are not available for automated canals that accurately predict water surface profiles when bore waves form. A good program will detect these conditions and either stop or alert the program user when a bore wave has formed during the computations.

2-8. Storage Considerations

As noted previously, the velocity of translatory waves is a function of canal water depth. Thus, operation of a canal system is enhanced when water depths are maximized. A more important consideration, however, is the change in storage relative to the change in canal flow. With a relatively small backwater effect in a canal, the difference between water surface profiles for different flows is almost rectangular as shown on figure 2-11. The shaded area shows the volume of water that must be added to increase the flow from one steady-state flow to another. A negligible backwater effect condition exists when the control structure spacing is large relative to the channel slope.

By increasing the number of check structures, backwater effect causes the difference between water surface profiles to become wedges which are nearly triangular. Figure 2-12 shows how this can greatly reduce the storage volume between two different flow rates.

In examining wedge storage, it is evident that check gate spacing has a significant effect on the volume of water required to change to a new water surface profile. Doubling the number of check structures in a canal will approximately halve the volume change needed to establish a new steady-state flow. This can be seen on figure 2-13 where the shaded area represents the reduction in wedge storage that would result from placing a check structure in the middle of a pool.

When flow at a canal headworks is changed, the time needed for the associated translatory wave front to reach any given point in the canal downstream can be determined from wave speed. However, the steady state will take longer to achieve. The time required for a new steady state flow to be established can be estimated as:

$$T_s = \frac{\Delta V}{\Delta Q}$$

where:

T_s = time to achieve new steady-state flow
ΔV = change in storage volume between previous steady state and new steady-state conditions
ΔQ = change in flow

Therefore, flow changes can be accomplished more quickly when check structure spacing, and hence, the wedge storage volume, is reduced.

DELIVERY CONCEPTS

2-9. General

The degree of freedom that water users have related to time and quantity of water they receive, defines the delivery concept. The principal delivery concepts are:

- Rotation—Users share a constant water supply, while cooperating with other water users to establish the time and quantity of delivery
- Scheduled—Advanced notification of delivery time and quantity is required, whereby water users are limited to contractual water allotment on a daily, monthly, and yearly basis
- Demand—Unrestricted use of the available water supply with limitations only on maximum flow rate and total allotment

The delivery concept establishes important operating criteria for the main canal system. Flow response characteristics in the canal system are different for each delivery concept. The rotation delivery concept

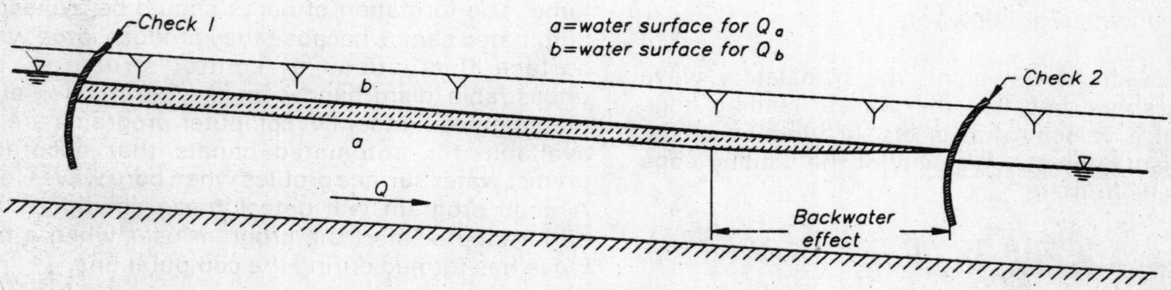

Figure 2-11. — Storage with negligible backwater effect.

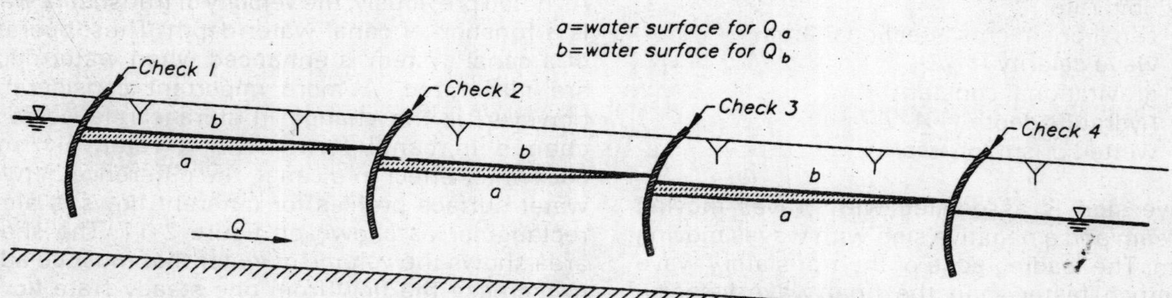

Figure 2-12. — Wedge storage effect.

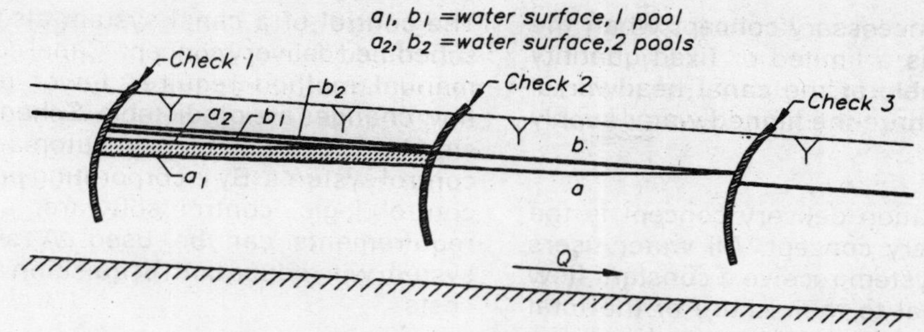

Figure 2-13. — Effect of added check structure.

offers the least demanding canal operation, because flow changes at the canalside turnouts are infrequent and water level fluctuations in the canal prism are minimal. More flow changes will be experienced with the scheduled delivery concept, which increases the complexity of canal operations. The predictability of these scheduled flow changes helps to minimize the resulting water level fluctuations. For the demand delivery concept, the main canal needs to respond immediately to unannounced and frequent flow changes at the canalside turnouts. Unannounced flow changes may cause relatively large water level fluctuations in the canal prism.

2-10. Rotation Delivery

The rotation delivery concept rotates a relatively constant canal turnout flow among a group of water users. The constant turnout flow is shared in time and quantity so that each water user receives a fair portion of the total diversion. Rotation delivery probably causes the most inconvenience to individual water users. They must cooperate with one another so that all receive their share of the constant available water supply.

Usually, farmers must rotate the specific timing of their irrigation schedule to improve the efficiency of water application to their crops. Last minute changes in the schedule are not allowed because they would have an adverse effect on the schedule of several other water users. It is a take-it-or-lose-it proposition. Water users will experience severe economic losses if they are deficient or uncooperative.

Even when maximum proficiency and cooperation are achieved, the rotation delivery concept causes inefficient irrigation water use. The crop area to be irrigated will be based upon the fixed delivery rate during the peak water use period. During offpeak periods, such as early and late in the irrigation season, more water than necessary may be available. Excessive irrigation during these periods causes additional surface runoff and drainage problems.

Inflexibility of rotation delivery creates other disadvantages for the water user, including:

- The set duration and frequency of delivery will be inappropriate for much of the service area, because varying soil conditions will cause different infiltration rates.
- Crops must be restricted to those that are well suited to the fixed frequency of irrigation.
- Disease and pests are more difficult to control.
- Crops such as vegetables, that require prompt irrigation when fields are dry, are difficult to manage.

The major advantage of the rotation delivery concept is the facility of main canal operation; i.e.,

- The operational difficulties and expenses are minimized.
- Flow changes are infrequent and usually small in magnitude from day to day.
- Water level fluctuations are minimal.
- Balanced operation is easily performed because the flow from canalside turnouts is constant and cannot be changed by the water users downstream.
- The main canal system can be operated by ditchriders (local manual control).
- Canal automation usually is not required because flow remains relatively constant for long periods of time.

Considerable cost savings can be realized in the construction of canal systems when the rotation delivery concept is employed:

- The main canal size can be optimized because the canal prism capacity is only large enough to convey the average flow during peak use periods.
- As only a portion of the users receive water at any one time, canal capacity can be set at about 40 percent of the total water user allotment.
- Even greater savings occur in constructing the distribution systems because rotating turnout deliveries (among water users) can considerably reduce the required maximum design capacity.

Rotation delivery is a necessary concept when the main canal system has a limited or fixed quantity of water supply available at the canal headworks. All water users must share the limited water supply on an equitable basis.

A variation of the rotation delivery concept is the continuous flow delivery concept. All water users in a continuous flow system receive a constant flow rate that is proportional to their share of the total flow. Continuous flow delivery has the advantages and disadvantages of rotation delivery except the water users have even less flexibility. Consequently, water users often make arrangements to rotate water use with neighbors.

2-11. Scheduled Delivery

The prevalent delivery concept in the United States is one in which water delivery is scheduled based upon advanced notification. Individual water users order water by specifying the time of delivery and the quantity of water they wish to receive. Then, individual water orders are compiled to predict the required canalside turnout flow changes. All turnout flows are added up to obtain the total flow schedule for the canal.

Scheduled delivery causes some inconvenience and possible economic losses to water users. One or more days in advance, water users must decide the quantity of water needed at their turnout; then, they must adhere to that announced schedule. The scheduled delivery may interfere with other unforeseen farm activities. If anticipated water requirements in the plant root zone have been underestimated, plants may become stressed or wilted. Overestimation of water quantity can result in wasted water, runoff and drainage problems, and other inefficiencies.

Inaccurate estimation of water needs will result in decreased crop production. Usually, canal operators hasten to accommodate users' last minute water schedule changes, particularly if a potential for crop loss is apparent.

A canal system is easier to operate when deliveries are scheduled in advance. Main canal flow changes are predictable and normally can be accomplished without major water level fluctuations. Changes in water depth are kept to a minimum by releasing flow into the canal before the actual time of turnout deliveries. Therefore, response time is accommodated in advance—diminishing the possibility of disrupting service to other canalside turnouts. An operation that minimizes water level fluctuations decreases the potential for canal lining and embankment failures.

The control of a canal system also is aided by the scheduled delivery concept. Canal control by the local manual method requires fewer ditchriders when flow changes are predictable. Scheduled delivery can simplify the design of local automatic or supervisory control systems. By incorporating prediction, simpler control logic, control software, and performance requirements can be used. A simplified control system will reduce both acquisition and maintenance costs.

The maximum conveyance capacity of a canal having scheduled delivery will typically need to be about 60 percent of the total individual water user demand. The individual water user's schedule can be adjusted to prevent the design capacity of the main canal from being exceeded. Also, the watermaster can require water users to reschedule their deliveries if the cumulated water orders exceed the daily, monthly, or yearly contractual allotments.

2-12. Demand Delivery

The demand delivery concept allows users to take water from the canal system whenever they need it. The quantity of water delivered is limited only by the maximum delivery flow and total allotment restrictions. This concept of delivery is normal for municipal water distribution pipe systems.

Demand delivery offers the maximum flexibility and convenience to the water user. Receiving water on demand also has economic value to the water user because delays or quantity restrictions are not involved. The value of water supplied for irrigation can be much higher when it is provided to farmers exactly as needed, rather than only as available. Maximum productivity can be achieved when water is applied at a rate to match field conditions and for the proper duration. This will permit earlier establishment of crops, shorten the irrigation season, reduce weed and disease problems, allow more effective fertilizer use, and result in larger crop yields. Precise application of irrigation water will decrease the total water use and reduce the need for subsurface drainage.

The provision of service to users on demand is both difficult and costly. Unrestricted delivery can create major operational problems for a main canal system because it is difficult for an open channel system to respond quickly to unannounced flow changes. Unlike a pressurized pipeline, a long period of time is required to achieve a balanced operation following a flow change in a canal system. This slow response time can make it impossible for canal operations to keep up with demand changes, which causes excessive depth fluctuations, wasted water, and disruption of service to other canalside turnouts.

Also, a large water supply must be available at the head of the canal system to provide on-demand service.

A canal system that delivers water on-demand may have high operation and maintenance costs. A more sophisticated and expensive control system will be required to provide the necessary response characteristics. Canal linings, embankments, and control system equipment may require more maintenance and repairs than a system using other delivery concepts because of more frequent depth and flow fluctuations.

A canal system that will provide on-demand delivery is expensive to construct. The canal prism needs to be sized to convey a larger peak demand because the number of users taking water simultaneously is without restriction. Typically, the canal design capacity will be about 80 percent of the total entitled flow to all water users, assuming a maximum of 80 percent of all users will want water at any particular time.

2-13. Comparison and Combination of Delivery Concepts

The following table summarizes the advantages and disadvantages of the three delivery concepts, and compares some of the major considerations relating to delivery alternatives.

Comparison of Delivery Concepts

Consideration	Rotation	Scheduled	Demand
User convenience	Poor	Good	Excellent
Irrigation flexibility	Poor	Moderate	Excellent
Crop yield	Low	Medium	High
Water use efficiency	Low	Medium	High
Ease of canal operation	Easy	Moderate	Difficult
Design capacity of canal as percent of total water user allotment*	~40%	~60%	~80%
Complexity of the control system	Simple	Moderate	Complex
Cost of canal system	Low	Medium	High

* This percentage is approximate, since these values are highly dependent on water use patterns [8].

Demand, scheduled, and rotation delivery concepts are not mutually exclusive. Most canal systems will have the operational flexibility to combine concepts to some extent. Historically, water service on Bureau water projects is accomplished by a combination of scheduled and rotation delivery. The delivery flows requested by water users are distributed by the operating personnel—as required. During peak demand season, on laterals which are pressed to maximum capacity, this may result in essentially rotation delivery. Combining the delivery concepts minimizes the construction, operation, and maintenance costs; the objectives of the canal system can be achieved in an efficient manner.

Demand delivery has not been commonly used as the primary concept of delivery on canal systems. However, in some cases, on-demand service has been permitted when the flow changes were small and could be accommodated within the unused flow capacity of the canal prism. Modern control equipment can make demand delivery a feasible concept. The long term benefits for the water users can justify increasing the capability of canal systems to accommodate demand delivery.

The delivery concept has a major impact on the overall operations of a canal system. It should be a primary consideration when evaluating alternative operations for either new or existing canals.

METHODS OF OPERATION

2-14. Pool Operation Alternatives

Several methods are available which can be used to convey water downstream through a series of canal pools. The methods of operation should not be confused with canal operation and control concepts as described in chapter 1. The method of operation—sometimes referred to as "method of pool operation"—determines how the water level varies in a canal pool to satisfy the operation concept.

A canal's recovery characteristics—the speed and manner in which the canal recovers to a new steady-state flow after a flow change—are dependent upon the method of pool operation.

The method of operation is based upon the location of the canal pool water surface pivot point. The pivot point is the location within a canal pool at which the depth remains constant while the water surface slope varies. The methods of operation as shown on figure 2-14 are:

1. Constant downstream depth—The pivot point is located at the downstream end of the canal pool (fig. 2-14a)
2. Constant upstream depth—The pivot point is located at the upstream end of the canal pool (fig. 2-14b)
3. Constant volume—The pivot point is located near the midpoint of the canal pool (fig. 2-14c)
4. Controlled volume—The pivot point can move within the canal pool (fig. 2-14d)

The basic method of operation should be identified for a canal system before control alternatives are

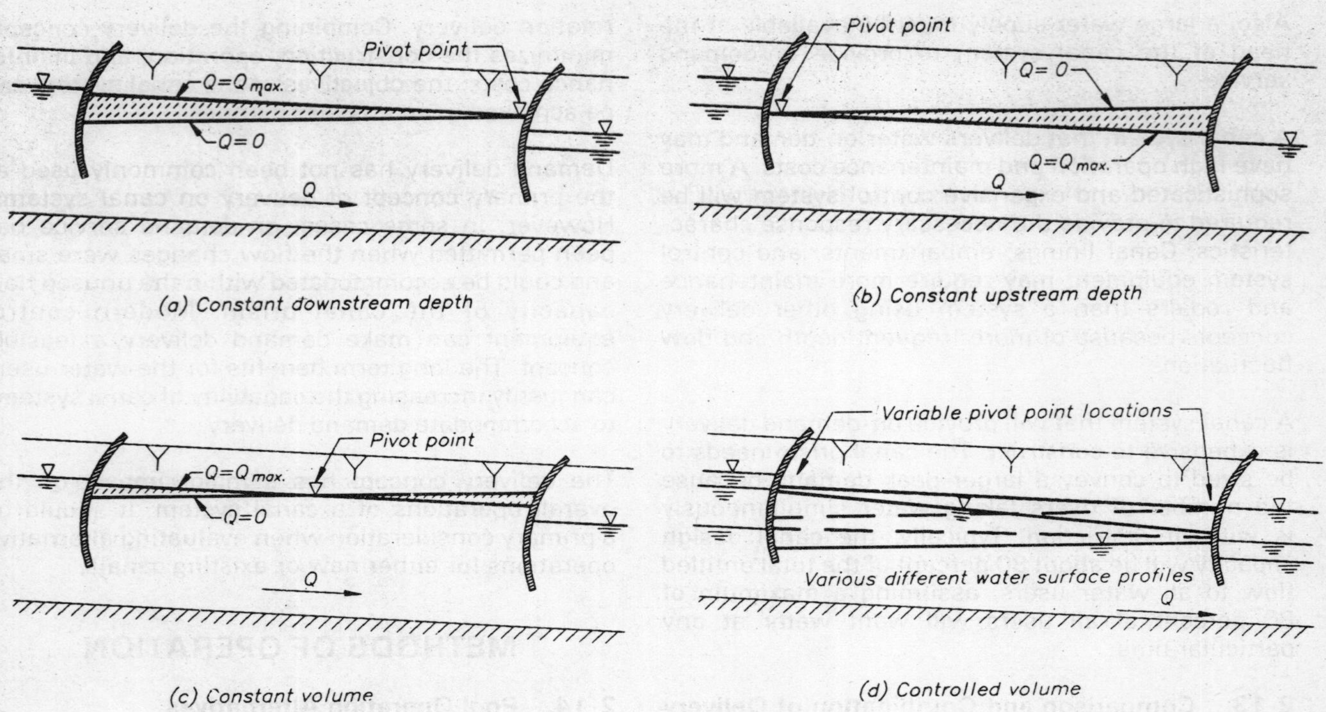

Figure 2-14. — Methods of operation.

evaluated. The location of the pivot point is particularly important when selecting a control method; i.e. local manual, local automatic, or supervisory control. Chapters 3 and 4 have information on control methods and their relation to methods of operation.

2-15. Constant Downstream Depth

Constant downstream depth method of operation—wherein the water depth at the downstream end of each canal pool remains relatively constant—is used in most canal systems. This method is associated with "conventional" operations and with local manual control, as discussed in chapters 1 and 2. The primary reason why this method is so prevalent is that a canal can be sized to convey the maximum steady flow; steady-state water depths should never exceed the normal depth for the design flow rate. The canal prism size and freeboard can be minimized, thus reducing construction costs.

With a constant downstream depth, major turnouts usually are located near the downstream end of the canal pools. This allows turnouts to be designed for a maximum and relatively constant depth in the canal, and also prevents problems in water delivery to users caused by low or fluctuating water depths. Wasteways, if any, also will be located at the downstream end of pools. Side-channel spillways and overflow weirs in the check structures can be set slightly above the maximum normal depth to prevent excessive depths.

When a constant depth is maintained at the downstream end of canal pools, the water surface profile will essentially pivot about this point as the canal flow changes (see fig. 2-14a). A storage wedge between different steady-state flow profiles is created. When flow increases, the water surface gradient and storage volume must also increase. Conversely, storage volume must decrease for a reduction in steady-state flow.

Because of these storage considerations, the constant downstream depth method of operation is particularly effective when combined with the upstream operation concept having a supply-oriented operation. A natural tendency exists for a flow change that originates at the upstream end of a pool to create the change in storage that is needed to keep the downstream pool depth constant.

Figure 2-15 shows how a flow change at the upstream end of a pool causes the pool storage to change in a manner compatible with the constant downstream depth method of operation. When pool inflow increases (fig. 2-15a), an additional volume of water enters the pool. This additional volume supplies the increase in storage required to achieve the higher surface gradient (fig. 2-15b). When the inflow decreases (fig. 2-15c), pool outflow will

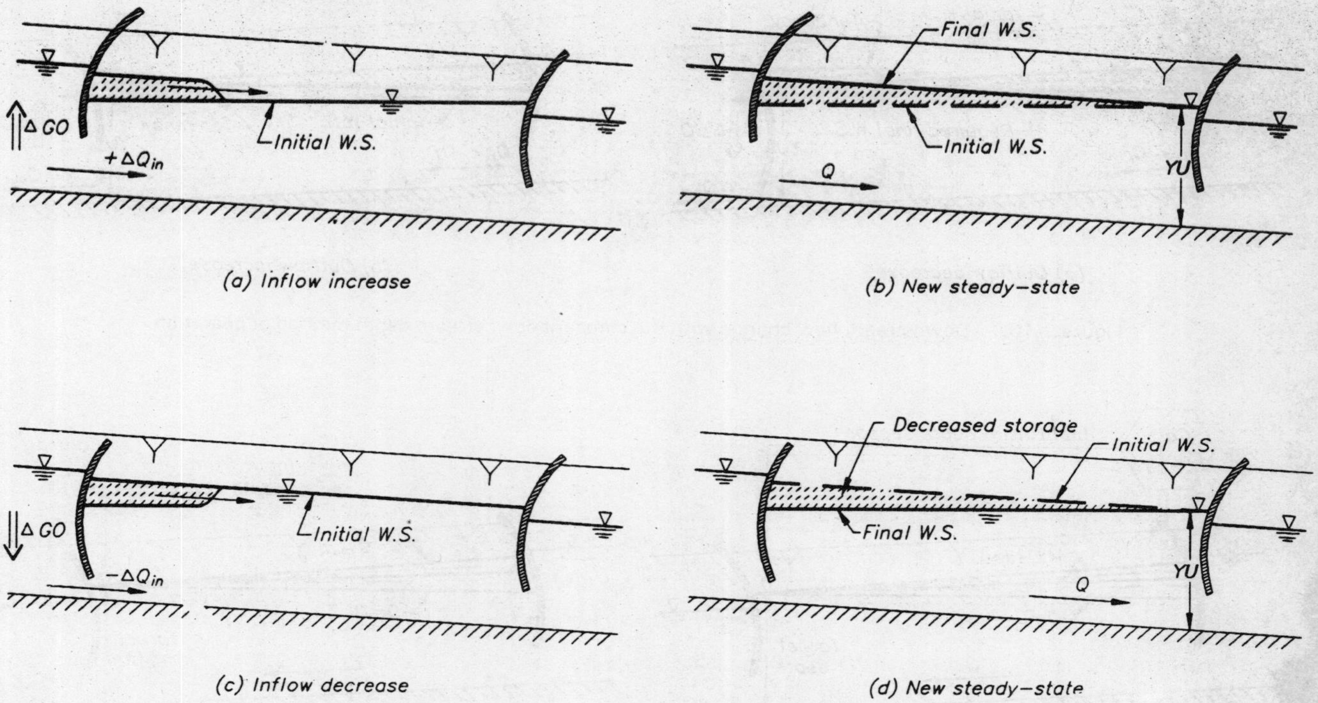

Figure 2-15. — Upstream flow change with the constant downstream depth method of operation.

temporarily exceed pool inflow. This will cause a decrease in pool volume, as required to lower the water surface to the desired final steady-state flow profile (fig. 2-15d).

Constant downstream depth method of operation has disadvantages, when combined with the downstream operation concept (demand-oriented operation), because pool storage must change oppositely to the natural tendency. As shown on figure 2-16, a flow change from the downstream end of a pool causes the depth to change in the wrong direction. When a decrease in pool outflow occurs (fig. 2-16a), the tendency is for pool storage to increase. To pivot the water surface about the downstream end, however, pool volume must decrease. The same problem will exist with an outflow increase (fig. 2-16b), as the tendency to reduce pool volume will be contrary to the requirement of increasing it.

To accomplish the required volume changes, inflow change at the pool's upstream end must overcompensate for the outflow change. Inflow must be changed by a greater amount than outflow until the new steady-state profile is achieved. Alternatively, if changes in demand can be predicted, inflow can be changed in advance of the outflow change.

Therefore, in a canal that operates with the constant downstream depth method, changes in demand are much more difficult to manage than changes in supply. The canal responds more easily and more quickly to supply-oriented operations than to demand-oriented operations. Flow changes that originate at the upstream end can be fairly large without creating problems. Flow changes originating from the downstream end must be relatively small and gradual to avoid excessive depth fluctuations. Often, anticipation is used to improve the system response to changes in demand by creating upstream flow changes in advance of demand changes.

Any method of control described in chapter 1 can be used to implement the constant downstream depth method of operation. Local manual control traditionally has been used, but automatic or supervisory control can successfully implement constant downstream depth method of operation. The target depth (setpoint), must be referenced at the water surface pivot point. The depth at this point can be controlled with either the upstream or downstream canal check gate as shown on figure 2-17. The upstream gate would be controlled to satisfy a downstream operation concept (fig. 2-17a), and the downstream gate would be controlled to satisfy an upstream operation concept (fig. 2-17b).

2-16. Constant Upstream Depth

With this method of operation, a constant upstream depth is maintained by pivoting the water surface at the upstream end of the canal pool as shown on

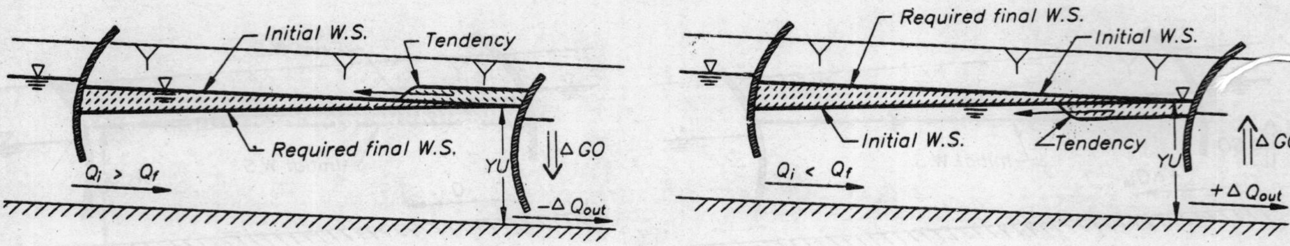

(a) Outflow decrease

(b) Outflow increase

Figure 2-16. — Downstream flow change with the constant downstream depth method of operation.

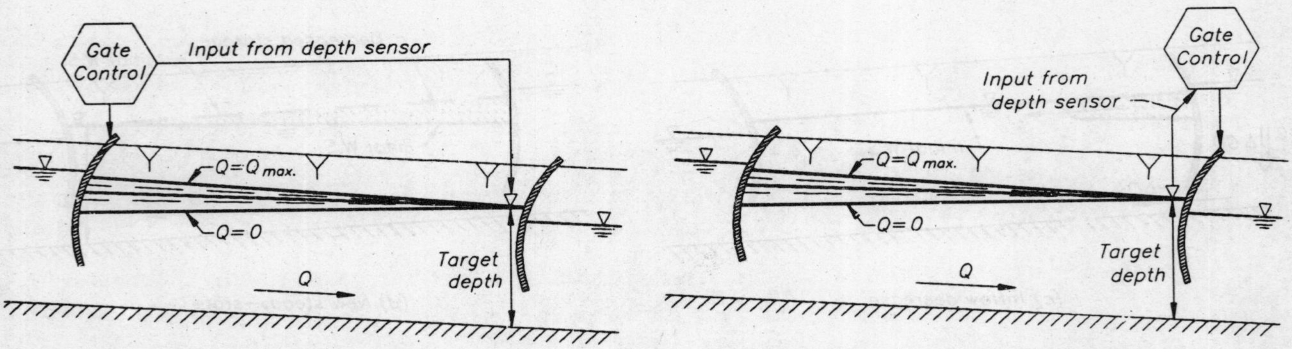

(a) Automatic control with upstream gate to satisfy downstream operation concept

(b) Automatic control with downstream gate to satisfy upstream operation concept

Figure 2-17. — Control alternatives for the constant downstream depth method of operation.

figure 2-18. The constant upstream depth method is sometimes called "level bank" operation, because canal banks must be horizontal to accommodate the zero-flow profile.

The construction of a level bank canal is the main drawback to this method. A level canal bank increases the cost of construction considerably, especially for concrete-lined canals. Most existing canals could not use level bank operation unless canal bank and lining were added to the downstream portion of each pool. Exceptions to this would be canals with little elevation drop between checks, or those operating at flows well below maximum flow capacity.

Turnouts can be located anywhere within a pool, as canal water depth should always be at least the full-flow normal depth. If constant head is required, however, turnouts should be at the upstream end of pools. Wasteways should be located similarly for best operation. If bypass weirs are to be included in check structures, the weir crest must be higher than the zero-flow depth at the downstream end of the pool.

The constant upstream depth method of operation is most effective when combined with the down-

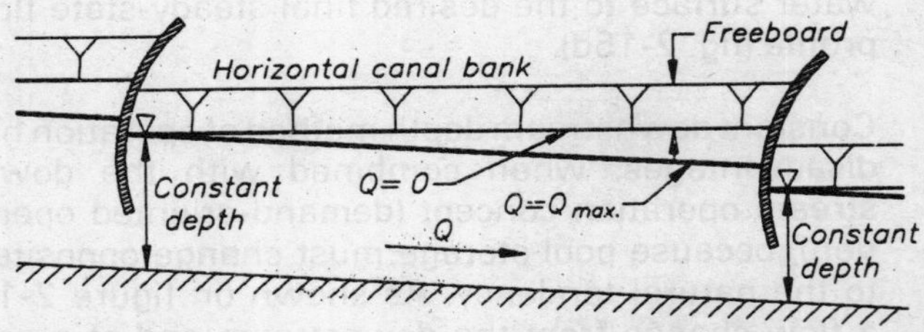

Figure 2-18. — Constant upstream depth method of operation.

stream operation concept (demand-oriented operation). Flow changes originating at the downstream end of the pool cause canal water depths to change in the direction needed to achieve new steady-state profiles.

Figure 2-19 illustrates the response of a level bank canal to changes in demand. A decrease in pool outflow will initially cause the downstream depth to rise (fig. 2-19a). This increase in pool storage will facilitate the change to a raised water surface profile, as required to pivot about the constant upstream depth. Conversely, an increase in outflow decreases the downstream depth towards the desired lower steady-state level (fig. 2-19b).

This excellent response to demand is the major advantage to level bank operation. Essentially,

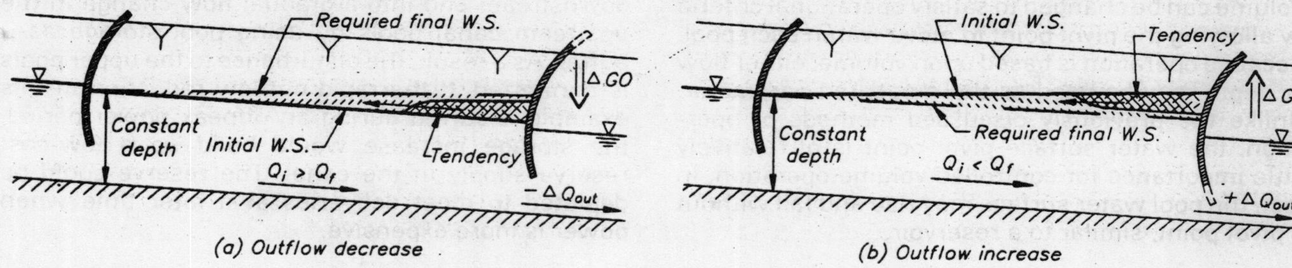

Figure 2-19. — Downstream flow change with the constant upstream depth method of operation.

regulatory storage is built into the canal. At low flows, the additional water stored in canal pools is readily available to supply increases in demand. At higher flows, storage volume is available above the water surface where excess canal flow can be stored in the event of a decrease in outflow. This allows the pool to react to demand, since the changes in storage within a pool can take place after downstream flow changes are made.

Level bank operation is inappropriate for supply-oriented canals. The operation would be inefficient and the additional expense for level bank construction would be unjustified.

Level bank operation can be accomplished successfully with any method of control described in chapter 1. Control can be based upon maintaining the target depth at the upstream pivot point; this allows the target to be located immediately downstream of the check gate structure being controlled. Automatic float-actuated gates use this principle to react to downstream demand [9, 10]. Better response to demand can be achieved by locating a depth sensor at the downstream end of the pool, to control the check gate at the upstream end. Control then must be based upon the variable depth at the downstream end of the pool.

2-17. Constant Volume

This method of operation is based upon maintaining a relatively constant water volume in each canal pool at all times. The water surface will pivot about a point near midpool as the flow changes from one steady-state to another. The constant volume method of operation is sometimes called "simultaneous operation", because the simultaneous gate operating technique often is used to keep the pool volume constant.

Storage wedges will exist on either side of the midpool pivot point as shown on figure 2-20. For any given flow change, volume change in each of these wedges is equal and opposite. When flow decreases, volume of water in the upstream wedge

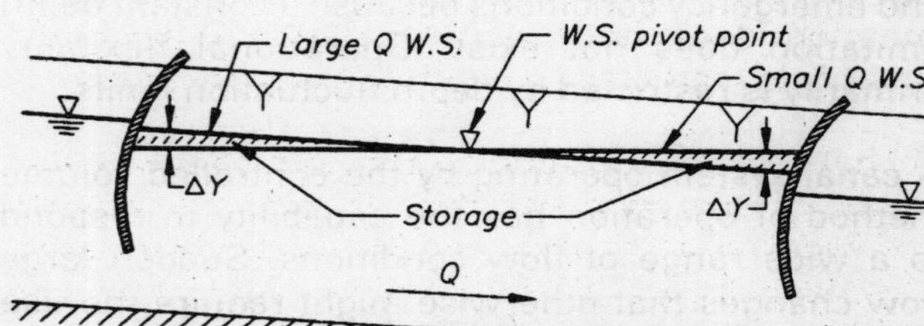

Figure 2-20. — Constant volume method of operation.

decreases and volume increases in the downstream wedge. When flow increases, the opposite occurs.

The main advantage of constant volume method of operation is the ability to quickly change flow conditions in the entire canal system. With the constant upstream depth and constant downstream depth methods, excessive time is required to either build up or deplete the storage in the entire canal system when changing the steady-state rate of flow. The constant volume method of operation avoids lengthy delays, because total volume of water in the canal system does not change significantly.

One disadvantage of constant volume method of operation is the additional canal bank and lining required at the downstream end of each pool, as compared to a conventional canal bank. However, the additional height required to accommodate the zero-flow water surface is only about one-half that required for level bank operation.

Another disadvantage involves the adjustment of check structures. In using local manual control, constant volume method of operation is quite difficult to accomplish. Local automatic control can be used, but the supervisory control method is best suited for the constant volume method of operation. Using supervisory control, all control structures can be adjusted simultaneously from a central location.

2-18. Controlled Volume

A canal system can be operated by managing the water volume contained in one or more canal pools.

Volume can be changed to satisfy operational criteria by allowing the pivot point to move within each pool. Because operation is based upon volume, either flow or depth may be used as the measured parameter. Unlike the previously discussed methods of operation, the water surface pivot point is of relatively little importance for controlled volume operation. In fact, the pool water surface may rise and fall without a pivot point, similar to a reservoir.

The controlled volume method of operation offers the most flexibility of any method of operation. Canal operation can adapt more easily to normal, abnormal, and emergency conditions because a constant depth limitation does not exist. Operational flexibility primarily is restricted by depth fluctuation limits.

A canal system operating by the controlled volume method of operation has the capability to respond to a wide range of flow conditions. Sudden large flow changes that otherwise might require the use of off-channel storage or wasteways often can be controlled successfully within the canal using the controlled volume method.

The controlled volume method of operation is particularly suitable to operate canal systems that include offpeak pumping considerations. (Electric power often is less expensive during offpeak periods, when power demand is low, than during on-peak periods, when power use is high.) Substantial savings in power costs can be achieved when pumping is maximized during the low-rate periods and minimized during high-rate periods. This can be accomplished by lowering water levels in canal pools during on-peak hours and raising water levels during offpeak hours.

An example of controlled volume method of operation is illustrated on figure 2-21. The example shows a rapid outflow reduction at the downstream end of the canal. The check gates upstream are all adjusted simultaneously to reduce the canal flow, but the amount of flow reduction is smaller at each successive check. Therefore, flow into each canal pool at the upstream gate is greater than outflow—increasing the volume of water in each pool. After the volume has increased by a desired amount, additional gate movements can match inflow and outflow to prevent excessive depths.

In creating the operations in this example, several aspects could be considered. Depth fluctuations can be managed to avoid rapid drawdown at the upstream ends of the pools without wasting water or exceeding the maximum depth allowed at the downstream ends. Limiting the drawdown can prevent damage to the canal lining. Another aspect might be to transform rapid flow change at the downstream end into a gradual flow change in the upstream canal pools by using pool storage as a buffer. As a result, the disturbance to the upper pools is minimized. If the rapid outflow reduction in this example occurred during an offpeak power period, the storage increase would build up a low-cost reserve supply in the canal. The reserve could be depleted to meet deliveries at a later time when power is more expensive.

One of the disadvantages of controlled volume method of operation is that it requires using the supervisory control method. Without computer assistance, the complexities of controlled volume operations would require frequent intervention by operations personnel. Computer-directed control would use specially developed software to control the check gates automatically—maintaining the desired volumes—without frequent operator intervention. Another disadvantage of controlled volume method of operation is the possible need for greater freeboard or a larger canal prism cross section.

CHECK GATE OPERATION

2-19. Check Structures

Canal control structures regulate the flow and depths of water. The most common type of canal control structure is the *check structure*. The name is derived from the function of checking, or slowing the flow velocity. A check structure causes the water surface to rise in the canal upstream of the structure. In some countries, check structures are called "cross regulators." A check structure can range in complexity from a simple fixed weir (in the channel) to a large structure containing many moveable gates [11, 12].

The canal check gate structure has become the dominant tool for implementing canal system operations. Because the check gate structure is the most common control structure, most of the examples in this chapter use check gates as the boundaries between canal pools. Typically, radial gates are used in large checks; vertical slide gates are more common in smaller check structures. The primary purposes of canal check gate structures are to (1) control water surface elevations, and (2) regulate the flow passing through the structure.

A schematic of a typical canal check gate illustrating steady-state free and submerged flow conditions is shown on figure 2-22. Usually, the check gate upstream depth, *YU*, is checked up to a desired level. Often, this level is prescribed to be the normal depth for the canal's maximum design discharge. A steady-state condition exists when depths and flow remain

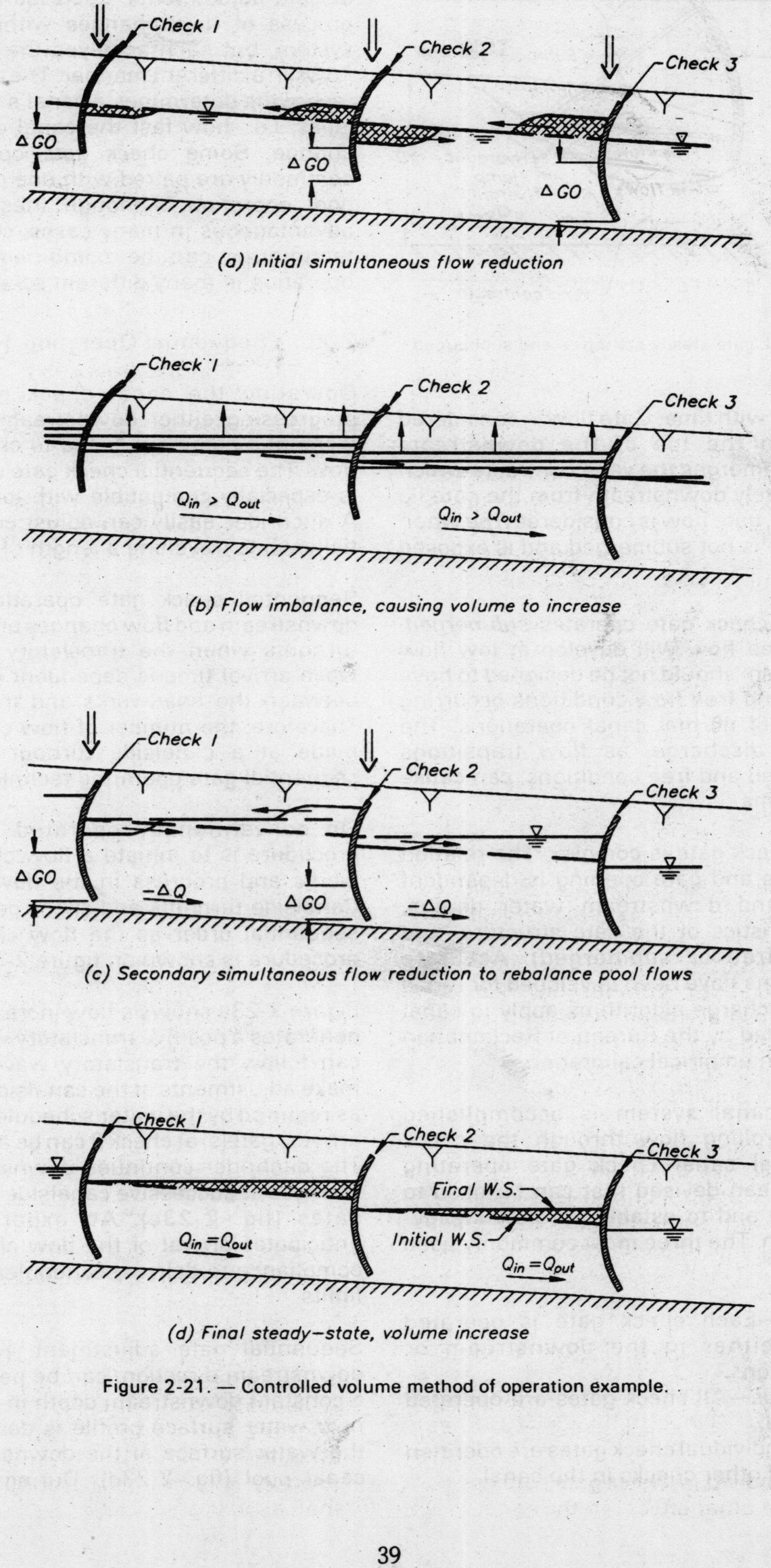

Figure 2-21. — Controlled volume method of operation example.

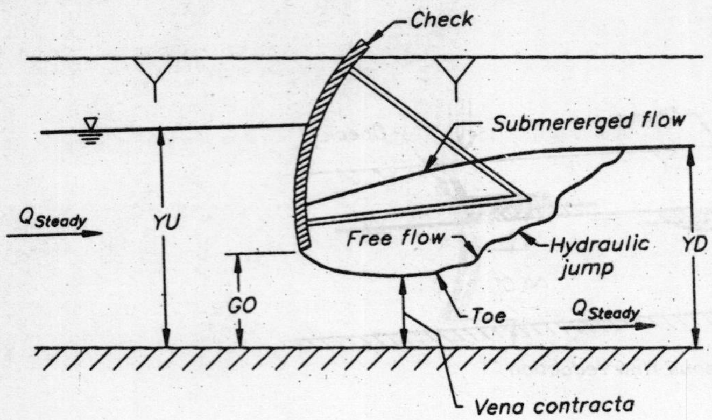

Figure 2-22. — Check gate steady-state free and submerged flow conditions.

relatively constant with time. Gate flow is considered *submerged* when the toe of the downstream hydraulic jump submerges the vena contracta which develops immediately downstream from the gate lip [5, 13]. Conversely, gate flow is considered *free* when the vena contracta is not submerged and is exposed to the atmosphere.

The typical canal check gate operates *submerged*. In some cases, free flow will develop at low flow rates. A canal system should not be designed to have both submerged and free flow conditions occurring within the range of normal canal operations. The rapid change in discharge, as flow transitions between submerged and free conditions, can cause operational problems.

Flow through a check gate is complex. The relation between discharge and gate opening is dependent upon upstream and downstream water depths, physical characteristics of the gate structure, and flow condition (free or submerged). Accurate discharge algorithms have been developed for radial gates [13]. The discharge algorithms apply to canal radial gates designed by the Bureau of Reclamation and are based upon empirical calibration.

Operation of a canal system is accomplished primarily by controlling flow through the check structures. Several canal check gate operating techniques have been devised that can be used to change canal flow and to establish a new steady-state flow condition. The three most commonly used techniques are:

1. Sequential—Each check gate is operated progressively either in the downstream or upstream direction.
2. Simultaneous—All check gates are operated at the same time.
3. Selected—Individual check gates are operated independently of other checks in the canal.

These three techniques differ primarily in the timing of gate adjustments. Each technique improves the process of flow changes within the entire canal system, but each achieves the desired depths and flows in a different manner. The check gate operating technique determines a canal's response characteristics; i.e., how fast the canal can respond to flow change. Some check gate operating techniques commonly are paired with one particular method of pool operation. Although these pairings may be advantageous in many cases, check gate operating techniques can be combined with methods of operation in many different schemes.

2-20. Sequential Operating Technique

Operating the canal check gates sequentially, progressing either downstream or upstream, is a technique commonly used to change canal system flow. The sequential check gate operating technique is especially compatible with local manual control. A ditchrider easily can adjust check gates sequentially while travelling a length of canal.

Sequential check gate operation transfers water downstream and flow changes are made at canalside turnouts when the translatory wave arrives. The wave arrival time is dependent on the canal length between the headworks and the turnout location. Therefore, the number of flow changes that can be made at a canalside turnout is limited for the sequential gate operating technique.

On conventionally operated canals, the basic procedure is to initiate a flow change at the headworks and progress in the downstream direction. Canalside turnouts and check gates are adjusted in sequential order as the flow change arrives. This procedure is shown on figure 2-23.

Figure 2-23a shows a flow increase (at check 1) that generates a positive translatory wave. The ditchrider can follow the translatory wave downstream and make adjustments at the canalside turnouts en route as required by the water schedule. Then, as the wave arrives, gate(s) at check 2 can be adjusted (fig. 2-23b). The ditchrider continues downstream making flow changes at successive canalside turnouts and check gates (fig. 2-23c). An experienced ditchrider anticipates arrival of the flow change and ensures compliance to delivery schedules within acceptable limits.

Sequential gate adjustment, progressing in the downstream direction, can be performed to achieve a constant downstream depth in the canal pool. The new water surface profile is developed by pivoting the water surface at the downstream end of each canal pool (fig. 2-23d). During the time interval

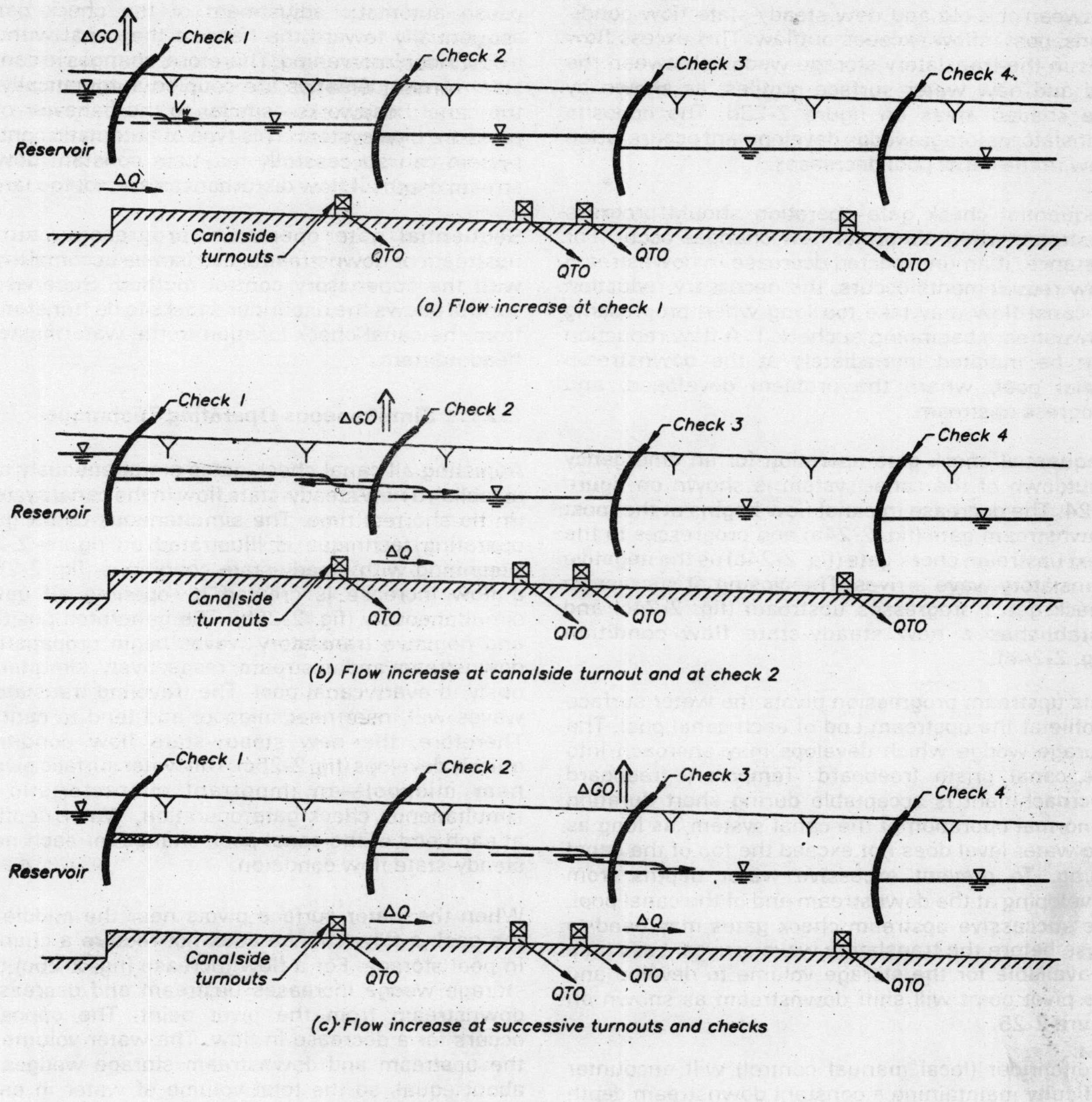

Figure 2-23. — Sequential gate operating technique progressing downstream.

between the old and new steady-state flow conditions, pool inflow exceeds outflow. This excess flow fills in the translatory storage wedges between the old and new water surface profiles, as shown by the shaded areas on figure 2-23d. The opposite translatory storage wedge development occurs when flow in the canal pool decreases.

Sequential check gate operation should progress upstream when abnormal flow changes occur. For instance, if an unexpected decrease in downstream flow requirements occurs, the necessary reduction in canal flow may take too long when progressing downstream beginning at check 1. A flow reduction can be initiated immediately at the downstream canal pool, where the problem developed, and progress upstream.

Sequential check gate operation for an emergency shutdown of the canal system is shown on figure 2-24. The decrease in canal flow begins at the most downstream gate (fig. 2-24a) and progresses to the next upstream check gate (fig. 2-24b) as the negative translatory wave arrives. The closing of successive check gates progresses upstream (fig. 2-24c) and establishes a new steady-state flow condition (fig. 2-24d).

This upstream progression pivots the water surface profile at the upstream end of each canal pool. The storage wedge which develops may encroach into the canal prism freeboard. Temporary freeboard encroachment is acceptable during short duration abnormal operation of the canal system, as long as the water level does not exceed the top of the canal lining. To prevent excessive water depths from developing at the downstream end of the canal pool, the successive upstream check gates may need to close before the translatory wave arrives. Less time is available for the storage volume to develop and the pivot point will shift downstream as shown on figure 2-25.

A ditchrider (local manual control) will encounter difficulty maintaining a constant downstream depth in canal pools with sequential check gate operation progressing upstream. The natural tendencies, of pool storage changes, are contrary to the downstream pivot point (see sec. 2-15). Therefore, only gradual flow changes can be accommodated without changing the depth at the downstream end of canal pools.

However, sequential check gate operation progressing upstream can be implemented with a local automatic control system. The automatic control system can be designed to respond to the downstream demands soon after they occur and to adjust the next upstream check gate. The resulting disturbances will continue to progress upstream and cause automatic adjustment of the check gates sequentially toward the head of the canal without a ditchrider intervening. Therefore, changes in canalside turnout demands are coupled automatically to the canal headworks—similar to the behavior of a pressure pipe system. This type of automatic control system can successfully maintain constant downstream depths if flow disturbances are not too large.

Sequential gate operation, progressing either upstream or downstream, also can be accomplished with the supervisory control method. Supervisory control allows the ditchrider's tasks to be transferred from the canal check location to the watermaster's headquarters.

2-21. Simultaneous Operating Technique

Adjusting all canal check gates simultaneously can establish a new steady-state flow in the canal system in the shortest time. The simultaneous check gate operating technique is illustrated on figure 2-26. Beginning with steady-state conditions (fig. 2-26a) a flow increase is created by opening all gates simultaneously (fig. 2-26b). The generated positive and negative translatory waves begin propagating downstream and upstream, respectively, simultaneously in every canal pool. The traveling translatory waves will meet near midpool and tend to cancel. Therefore, the new steady-state flow condition quickly develops (fig. 2-26c). The water surface pivots near midpool—an important characteristic of simultaneous check gate operation. Water depths, at each end of the canal pool, change for each new steady-state flow condition.

When the water surface pivots near the middle of the pool, a flow change does not require a change in pool storage. For a flow increase (fig. 2-26c) the storage wedge increases upstream and decreases downstream from the pivot point. The opposite occurs for a decrease in flow. The water volume in the upstream and downstream storage wedges is about equal, so the total volume of water in each canal pool will remain relatively constant for all steady-state flow conditions.

The depth fluctuations associated with the simultaneous gate operating technique must be scrutinized. Water level decreases should not exceed allowable drawdown rates for normal operation. Excessive drawdown can result in canal lining failure. During normal operation, increases in water depth should not be allowed to encroach into the freeboard for a sustained time. The canal bank and lining need to be high enough to accommodate the depth increases that are caused by simultaneous gate operations. The canal may have to be operated at lower than normal steady-state depths so that

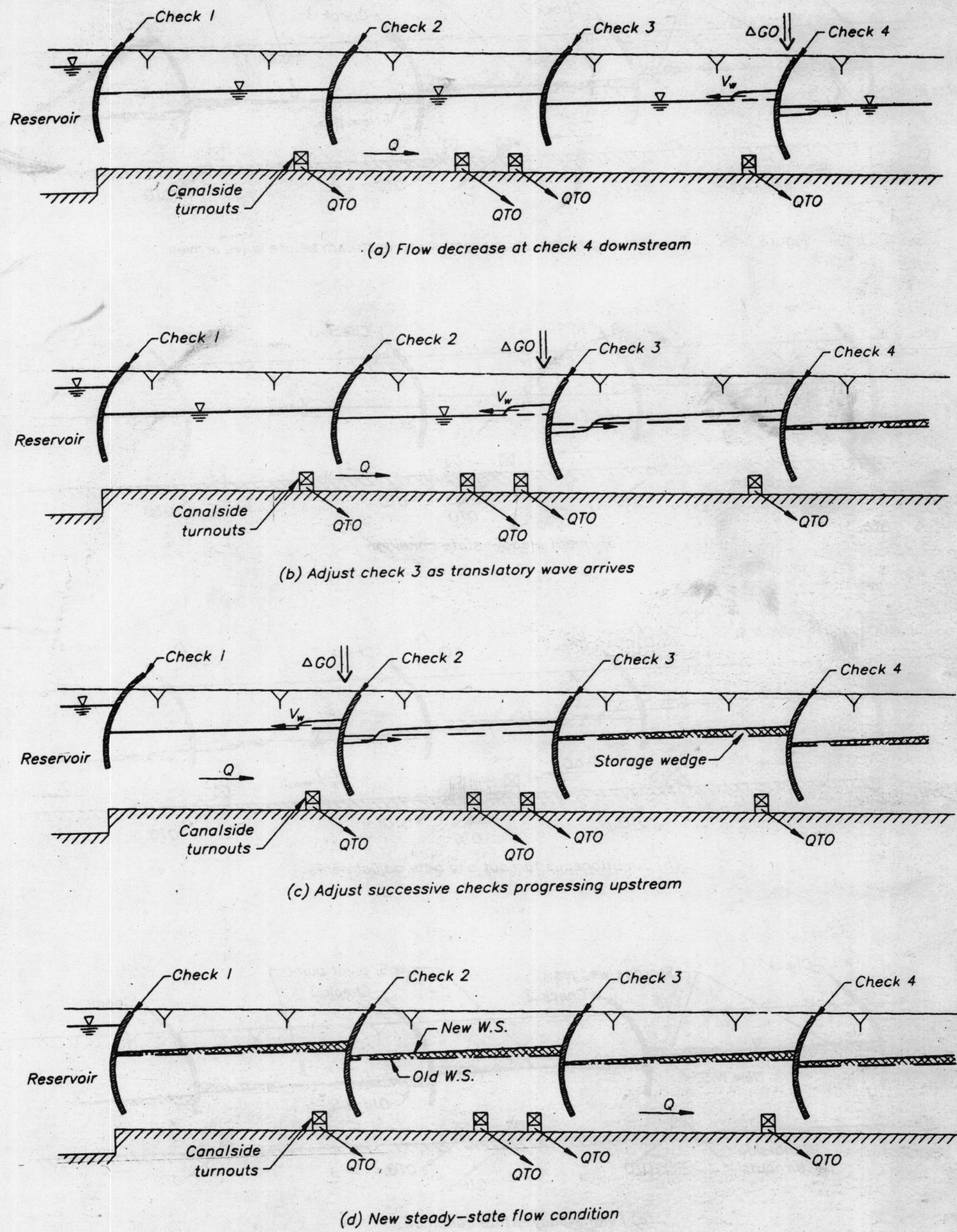

(a) Flow decrease at check 4 downstream

(b) Adjust check 3 as translatory wave arrives

(c) Adjust successive checks progressing upstream

(d) New steady-state flow condition

Figure 2-24. — Sequential gate operating technique progressing upstream.

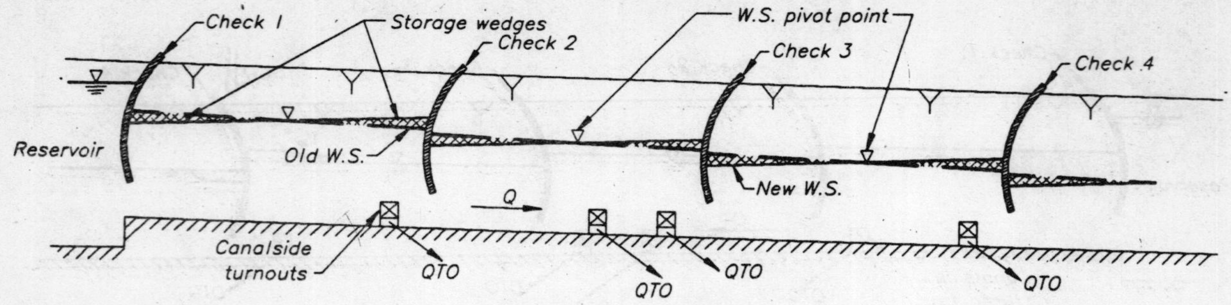

Figure 2-25. — Sequential gate operation progressing upstream before wave arrives.

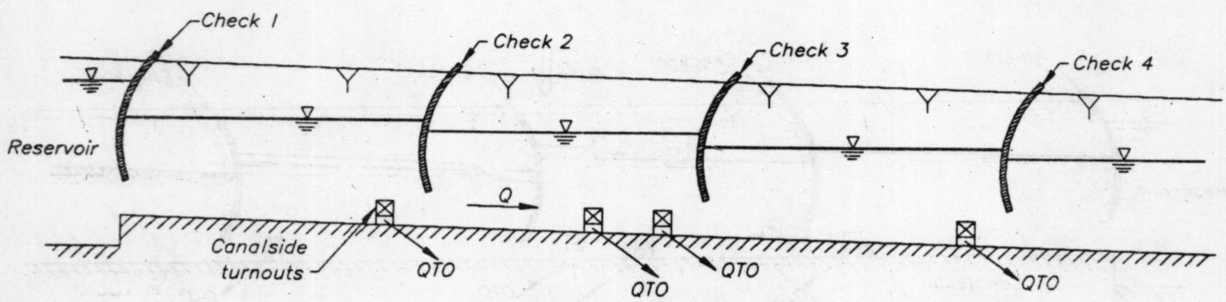

(a) Initial steady–state condition

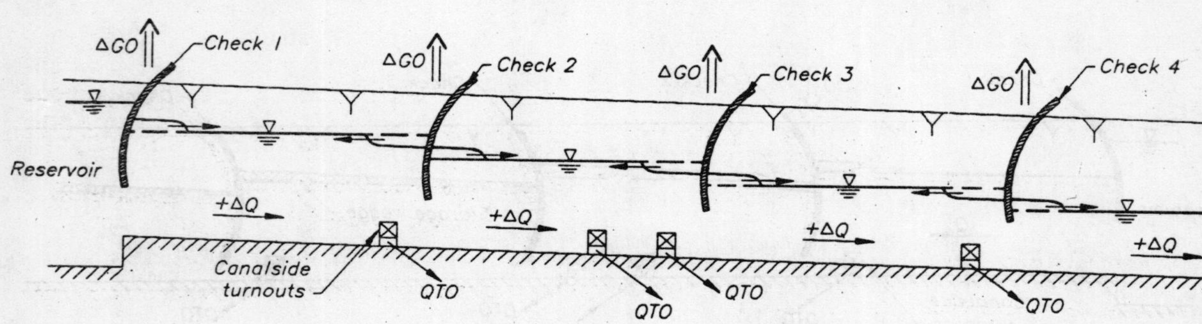

(b) Simultaneous turnout and gate adjustments

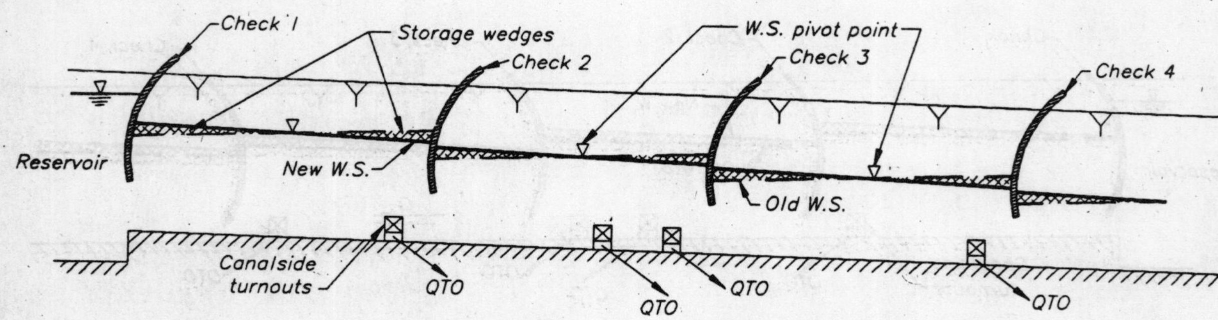

(c) Final steady–state condition

Figure 2-26. — Simultaneous gate operation with midpool pivot point.

sufficient freeboard is available for flow changes. Therefore, as steady-state flow approaches canal maximum design flow, the capability to make large flow changes decreases.

Simultaneous closure of all check gates is ideal for an emergency shutdown of the canal system. During an emergency, encroachment into the freeboard is permitted for a short time. Usually, a sudden simultaneous shutdown will minimize this encroachment compared with other check gate operating techniques which react more slowly to an emergency.

When using the simultaneous check gate operating technique, it is possible to shift the water surface pivot point either upstream or downstream before a new steady-state flow develops. A two-step adjustment of all check gates is required, with a time lapse between initial and final gate operations. The initial flow change creates an unbalanced flow condition in each canal pool to increase or decrease the storage wedge water volume. The secondary flow change then again balances the inflow and outflow after the desired pool volume change has been achieved. The secondary flow change may require more than one gate adjustment to achieve accurate flow balances.

For example, figure 2-27 illustrates how the pivot point of the water surface can be shifted from the midpoint to the downstream end of a canal pool during an increase in flow. The initial simultaneous gate operation (fig. 2-27a) includes an additional increment of gate opening, ΔGO, at all the canal check gates. This additional gate opening increment is always larger at the upstream check gate—creating an unbalanced flow into the canal pool. Therefore, Q_{in} is greater than QTO plus Q_{out}.

The intermediate water surface pivot point will occur near the midpoint of the canal pool as the generated translatory waves meet (fig. 2-27b). However, flow mismatch will increase the translatory storage until the pivot point has been shifted to the downstream end of the canal pool (fig. 2-27c). At this time, all check gates are simultaneously closed by the increment ΔGO so that inflow once again matches outflow (fig. 2-27d).

The additional increment included in the initial gate movement, ΔGO, can be selected to increase or decrease the storage. Thereby, the pivot point can be deliberately shifted upstream or downstream by the two-step simultaneous gate operating technique regardless of flow change (increase or decrease). Water volume in each canal pool will vary, if desired, for each steady-state flow condition. Two-step simultaneous check gate operation is one technique to enact the controlled volume method of operation.

Simultaneous gate operations cannot be accomplished by local manual control unless ditchriders are stationed at every check structure so they can communicate with each other when flow changes are to be initiated. Realistically, the simultaneous gate operating technique can be accomplished only

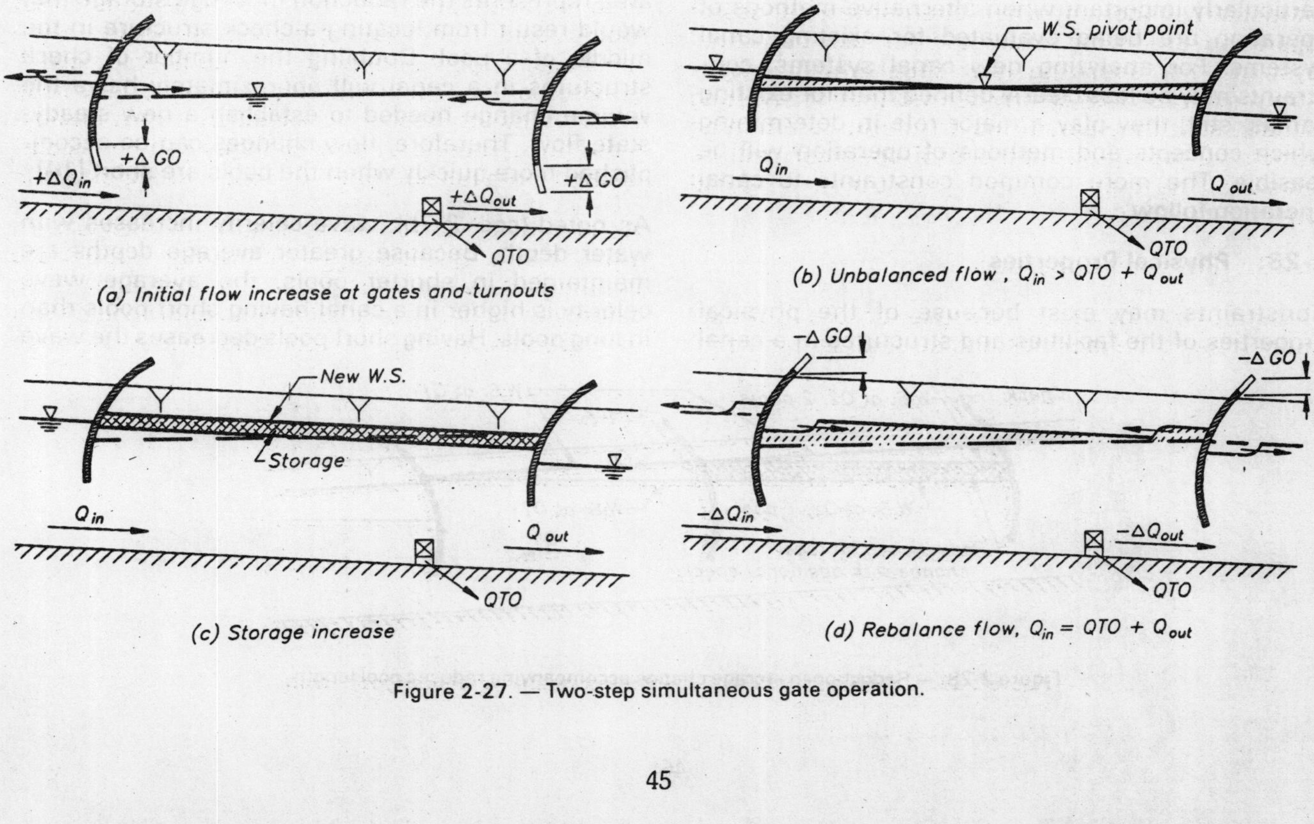

Figure 2-27. — Two-step simultaneous gate operation.

by using the supervisory control method. New gate openings can be transmitted rapidly from a central location to remote terminal units at all check structures so that all gates are adjusted almost simultaneously.

2-22. Selected Operating Technique

The selected gate operating technique is commonly used to make minor flow adjustments which do not necessarily affect the entire canal. Selected gate adjustments often are made to balance the operation of the canal system. Certain check gates may require an adjustment to maintain desired water depths in adjacent pools.

At times, water transfer may be necessary between surplus and deficient canal pools. The transfer can progress either downstream or upstream. If the water transfer involves more than one check gate structure in succession, either sequential or simultaneous gate operation could be used.

Selected gate operation permits adjusting water depths and flows within the canal system without having to adjust the headworks inflow. This type of operation is especially advantageous when the headworks consists of a pumping plant with a minimum flow increment determined by pump unit capacity.

CONSTRAINTS

The alternatives available for operating a canal system are limited by constraints. Constraints are particularly important when alternative methods of operation are being evaluated for existing canal systems. For analyzing new canal systems, constraints may be less clearly defined than for existing canals; still, they play a major role in determining which concepts and methods of operation will be feasible. The more common constraints to canal operation follow.

2-23. Physical Properties

Constraints may exist because of the physical properties of the facilities and structures in a canal system. The following paragraphs discuss some of the constraints associated with physical (structural) features of a canal system.

a. Pool length.—The distance between check structures has a large impact on the ability to control flow in a canal. If canal pools are long, initiating flow changes without causing large depth fluctuations will be difficult. The vertical drop between checks will be particularly important in this regard. The depth fluctuation between maximum design flow profile and zero-flow profile is proportional to the invert elevation change in the pool. The exact amount of fluctuation will vary with the method of operation employed.

During operations, pool length influences the responsiveness of a canal system. A canal with shorter pools is easier to control because it will respond to flow changes and achieve a new steady state more quickly than long pools. When constant upstream depth or constant downstream depth methods of operation are used, pool length has a significant effect on the volume of water required to change to a new water surface profile.

A canal that operates with the constant downstream depth method is shown on figure 2-28. Water surface profiles are shown for two different flow rates, Q1 and Q2, both with and without the middle check structure. The water surface for Q1 is not changed significantly by the check structures, because Q1 is near canal maximum flow capacity. But, at the lower flow, Q2, the water surface is much higher with the middle check than without it. The shaded area represents the reduction in wedge storage that would result from locating a check structure in the middle of a pool. Doubling the number of check structures in a canal will approximately halve the volume change needed to establish a new steady-state flow. Therefore, flow changes can be accomplished more quickly when the pools are short [14].

As noted (sec. 2-7b), wave celerity increases with water depth. Because greater average depths are maintained in shorter pools, the average wave celerity is higher in a canal having short pools than in long pools. Having short pools decreases the wave

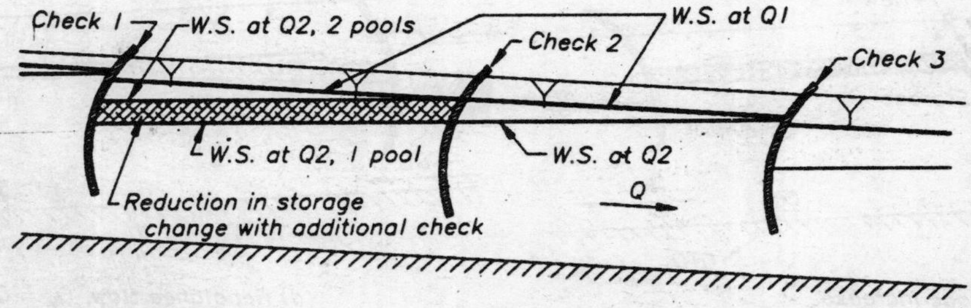

Figure 2-28. — Reduction in storage change accompanying reduced pool length.

travel time in the canal and improves response to flow change.

b. **Freeboard.**—Canals are constructed with some additional lining and bank height (freeboard) above the normal maximum water surface level. Freeboard is designed to protect the canal bank from wind generated waves on the water surface and from storm runoff inflows. Also, freeboard provides additional safety when depths accidentally exceed maximum normal levels. If water overtops the freeboard, it can damage the canal and flood adjoining areas.

For new canals, maximum water levels should be determined through operation studies before final design. Then, freeboard is added above the maximum levels to accommodate wind waves. Once lining and bank heights are fixed, canal operations must limit water levels below the maximum at all points throughout the canal system.

The constant upstream depth method of operation (level bank operation) requires horizontal canal banks because the water surface will be level at zero-flow. The constant volume method of operation may require a level bank in the downstream half of each pool and a prismatic channel in the upstream half. In most existing canals, freeboard constraint prevents using these methods of operation unless the canal is modified.

In addition to steady flow profiles of the different methods of operation, waves created by unsteady flows must remain below the canal lining height. Flow changes must be made in a manner slow enough so as not to overtop the lining at any location. The amount of freeboard is one factor that determines how rapidly the flow can be changed.

c. **Drawdown.**—Drawdown is the rate of depth reduction at any point in the canal. Drawdown criteria is one of the most restrictive constraints upon canal operation. Rapid increases in water depth are seldom a problem—unless the maximum depth is exceeded. However, rapid decreases can damage the canal even when depths remain within an acceptable range.

Damage will result if hydrostatic forces underneath or behind the canal lining material exceed those forces on the inside of the lining by a large enough magnitude. The difference in hydrostatic pressure required to damage any particular canal is a function of the strength and weight of the lining material and the shape of the canal prism. Sufficient force on the lining's backside will cause buckling, cracking, and bulges in the lining.

Whenever canal embankment soil is saturated with water, hydrostatic pressure will be exerted on the backside of the canal lining. The elevation of saturated soil (called the phreatic line) often will be slightly below the canal water level as shown on figure 2-29. The amount of canal embankment saturation varies with the porosity of the canal lining and the permeability of the embankment material. Sometimes, drains are placed behind the canal lining to remove water from the soil and reduce the back pressure. The embankment hydrostatic pressure will vary from one point to another along the canal length, but normally the canal's internal water pressure will be greater than or equal to the back pressure.

When canal lining is relatively impervious and soil behind the lining drains slowly, water level in the soil will not drop rapidly. If canal water level is reduced faster than the embankment saturation level, then the net hydrostatic pressure on the lining

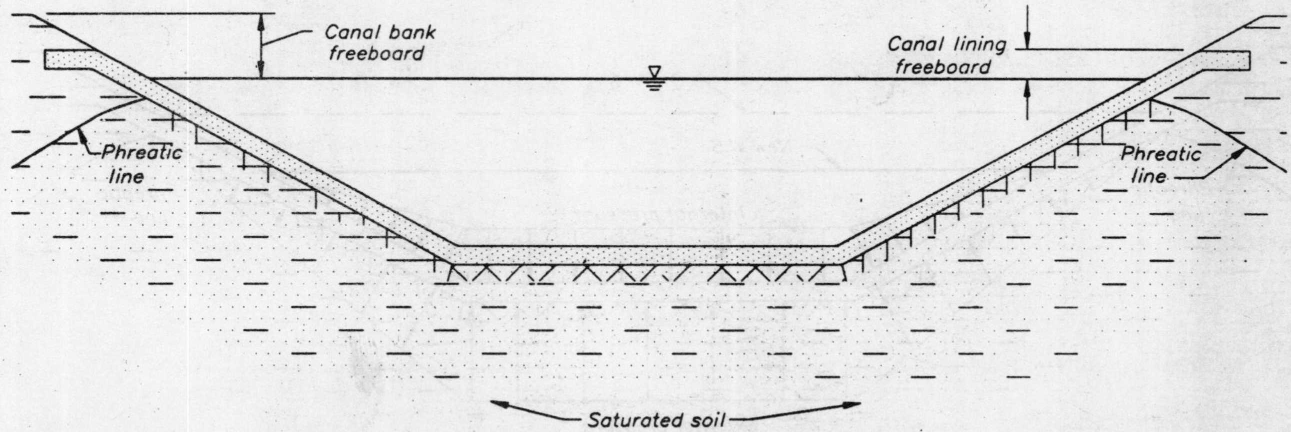

Figure 2-29. — Typical soil saturation behind canal lining.

becomes adverse as shown on figure 2-30. The weight and strength of the lining will withstand only a limited amount of adverse pressure before it is damaged.

A maximum acceptable drawdown rate should be established for each canal. Typical drawdown rates permitted in Bureau concrete-lined canals having 1.5:1 side slopes (horizontal:vertical) are 0.5 foot (150 mm) during any 1 hour and 1 foot (300 mm) during any 24-hour period. These rates are acceptable in most cases for earth-lined and membrane-lined canals. In many instances, higher drawdown rates may prove to be acceptable during operations. More rapid drawdown may be satisfactory in unlined canals, in areas of well drained soil, in sections built on fill, and in canals with heavy lining or having drains behind the lining. Usually, the normal drawdown criteria is relaxed for abnormal and emergency operations which occur infrequently.

d. Pumping plants.—A pumping plant can be a major constraint to canal operations. A canal supplied by pumps or having inline relift pumps to convey water to higher elevations is totally dependent upon pump operation. Equipment failure or power loss, at a pumping plant, can suddenly stop the entire canal flow. Storage, wasteways, and the control system must be designed appropriately.

A canal supplied by a headworks pumping plant can dewater when a power failure occurs. An automated canal using the downstream control concept would be prone to this problem; therefore, measures must be taken to prevent the canal from dewatering when supply stops. The upstream control concept must be used so that the system reacts properly to a power failure at the headworks pumping plant.

Considering the distance between control structures, pumps affect operations similarly to check structures. Unlike check gate structures, however, pumps significantly reduce the flexibility of flow adjustment. Unless pumps are equipped with variable-speed motor controllers, flow changes from a pumping plant are adjusted in steps based upon the capacity of each pumping unit. Operational flexibility also will be limited based upon criteria used for the frequency of pump starts and stops.

A constant speed pump must be either On or Off—producing full-flow or no-flow. Therefore, the smallest increment (of flow change) is the flow produced by the smallest unit in the pumping plant. Almost always, matching the exact supply to the required demand is impossible and regulatory storage is required to absorb flow imbalances. This storage often can be provided in the conveyance channel, but the operations of the canal will be influenced accordingly. With only a few pumps in a pumping plant, the constraint to operating a canal system is significant because the pump capacity increments are large relative to total canal flow.

Canalside pumping plants take water from the canal and pump it into a pipeline distribution system. Canalside pumping plants have similar characteristics to inline pumping plants, but usually on a smaller scale. Pumped turnouts can have the same problems with sudden flow change as inline pumping plants.

Sometimes, storage is required to insulate the main canal from a canalside pumping plant. A small forebay reservoir can be built to supply the pumping plant, thereby allowing relatively steady flow through a gravity turnout from the main canal into the forebay reservoir. If the pumping plant takes water directly from the canal, the canal operation method and the control system must have capability to respond quickly when power failure occurs.

Some newer canal systems have automatically controlled canal-side pumping plants that respond

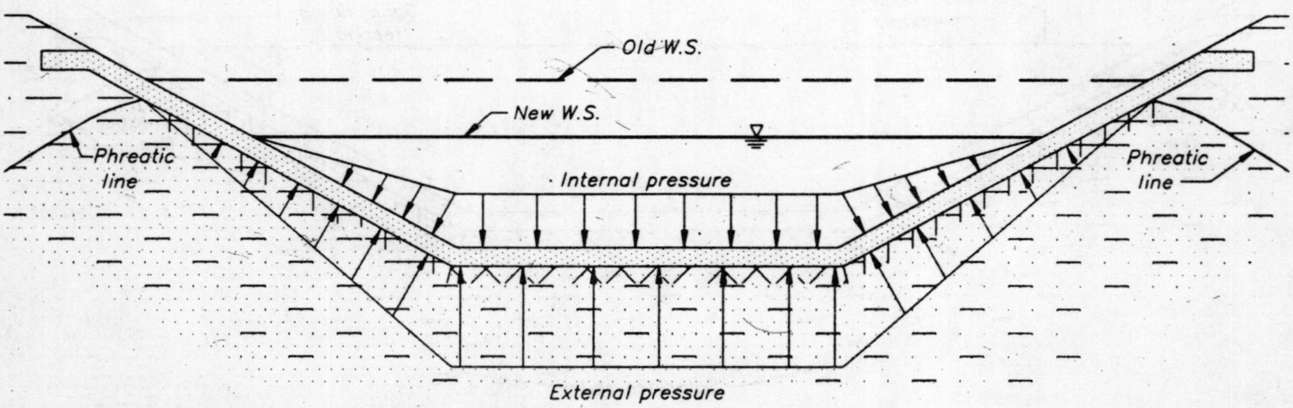

Figure 2-30. — Adverse hydrostatic pressure following rapid drawdown.

to downstream demand in the pipeline distribution system. These pumping plants can create large, unpredictable turnout flow changes during normal operations. A canal serving this type of delivery must have excellent response characteristics.

e. Turnouts.—The location, type, and size of turnouts (from the canal) are constraints on the operation of a canal. Water delivery to users depends on every turnout receiving an adequate supply from the canal. To make deliveries, all types of turnouts require a certain minimum water depth in the canal at the turnout location. Low water levels must be prevented at turnout locations to avoid insufficient water delivery.

At gravity turnouts, a relatively constant depth in the canal is required to provide steady turnout flow. The method of operation must be compatible with the location of gravity turnouts to satisfy the constant depth requirement. Check structures often are located just downstream of large gravity turnouts so as to maintain a constant depth at the turnout. With level-bank operation, gravity turnouts should be located near the upstream ends of the canal pools.

Pumped turnouts are less sensitive to the canal water depth. However, a minimum water level must be maintained for pumps to operate. Above the minimum level, fluctuations in canal water surface do not significantly affect the turnout delivery. Pumped turnouts usually can be located anywhere along the canal.

Pumped turnouts can affect operations because of the sudden flow changes they impose on the canal system. Normal flow changes, from pumps starting and stopping, may result in larger increments of flow change than normally would exist with a gated turnout.

Turnout size is important mostly for the amount of flow change which can occur in the canal due to turnout flow fluctuations. Turnouts that take a large percentage of the canal flow can affect operations significantly and must be specifically studied in the operation plans.

f. Check structures.—Most check structures contain one or more moveable gates; some use weirs or stoplogs to make flow adjustments. The adjustment increment for different types of check structures influences operations. Usually, gates can be adjusted in small increments, thereby offering a high level of operational flexibility. The adjustment of weirs and stoplogs typically can be made only in large steps, if at all, which can limit the operations. Additional limitation can stem from the frequency of adjustments or total number of adjustments permitted per day, depending on the gate type at the structure and the method used to control flow.

Many check gate structures contain sidebay overflow weirs—also called wing walls, overflow walls or bypass weirs—to pass flow around the gate(s) when the depth upstream of the check is too high. Often, it is beneficial to be able to pass excess flows downstream to prevent excessive depths in the canal upstream of the check. When the downstream canal has adequate flow capacity, excess water can be effectively conveyed to a wasteway or storage at points downstream. Some canals can be operated to pass some of the flow over the sidebay weirs during normal operations, which reduces the depth fluctuations upstream of the check structure and requires fewer gate adjustments.

But sidebay weir overflow can create problems downstream in some cases. The lower canal pools usually have a smaller flow capacity, so they may be unable to deal with additional flow. The crest of a sidebay overflow weir is usually only 0.1 to 0.2 foot (30 to 60 mm) above the maximum normal depth in the canal. This small weir freeboard may restrict the size of sudden flow changes at the check structure for normal operating conditions. Flow changes may have to be gradual to prevent sidebay overflow spill, although small spills usually would not create problems.

Sidebay overflow weirs also preclude using some methods of operation that require a greater than normal depth at the downstream end of the canal pool, such as the constant upstream depth and the constant volume methods of operation.

g. Drains.—Without underdrains, operational flexibility may be limited. Underdrains are used to drain the water from the soil around the canal prism to relieve hydrostatic pressure underneath or behind the canal lining. Drainage behind the lining affects the ability to reduce water depth in the canal without damaging the lining. If soil behind the lining is well drained, the rate of drawdown in the canal can be increased. Drains should be installed in areas of naturally high water tables, where the soil is particularly impervious, or where large drawdown in the canal level is anticipated. Since depth fluctuations tend to be concentrated in particular areas for each method of operation, soil drainage in these areas will be critical.

Drain inlet structures allow storm runoff to drain into the canal from adjacent terrain. Inflow from drain inlets may create problems of excess canal flow. Canal operations should be designed to accommodate excess flow supplied from drain inlets.

h. Closed conduits.—Canal systems sometimes contain segments wherein flow is conveyed through a closed conduit, as dictated by the natural terrain. If flow in a closed segment exhibits a free water surface (open-channel flow), the effects on canal system operations will be minor. Changes in channel properties will affect the hydraulics but should not present any significant operational problems. A closed conduit flowing full will have a larger impact on operations. In some cases, the effects of flow changes may be amplified, resulting in greater depth changes in canal sections adjacent to the closed conduits. In other cases, conduits may reflect or dampen hydraulic transients.

Inverted siphons often are used to convey canal flow underneath obstructions, such as natural drainage channels, where an open channel cannot cross. Siphons usually flow full (pressure flow) for most of their length. The siphon's invert elevation drop, between the upstream and downstream ends, usually is designed to be equal to or slightly greater than the total head loss at maximum design flow. Therefore, depth will be nearly the same in the channels on either side of the siphon at design flow. Figure 2-31a shows a profile of an inverted siphon at design flow conditions.

Because siphon head loss is proportional to the flow squared, head loss at low flows can be substantially smaller than the invert elevation drop. Therefore, at low flows, the water depth may change significantly from the channel upstream of the siphon to the channel downstream. Figure 2-31b shows the siphon profile at 50 percent design flow. The large reduction in depth results from the siphon head loss being changed by a factor of four between these two cases. Canal operations may need to be restricted to prevent exceeding drawdown criteria in the channel upstream of a siphon, unless the siphon has a gated entrance. When the elevation drop across the siphon is large, the possibility exists for entirely dewatering the upstream channel at zero-flow (fig. 2-31c).

In addition to drawdown problems, the volume of water depleted from the upstream channel during a flow reduction can create operational problems downstream. Excess water from the pool upstream of the siphon must be transferred downstream to be stored or wasted elsewhere in the canal system. If the upstream pool is long, a significant volume of excess water is involved.

Closed conduit segments that substantially constrict the flow area from an open channel section can partially reflect the waves caused by sudden flow changes. Wave reflection is especially important in channels near pumping plants or powerplants where power failures and load rejections can cause sudden flow rejection. The sudden flow changes cause bore waves that may be reflected and amplified as they reach the restricted section, thus increasing the maximum depths attained in the open channel.

2-24. Storage

a. In-channel storage.—The canal prism size will significantly affect operational flexibility to change the canal system flow. When canal flow is near the capacity of the conveyance channel, flow must be changed gradually to avoid creating unacceptable depths in the canal. When flow is substantially below maximum channel capacity, flow change is easier to accomplish.

Figure 2-32 shows a profile of a segment of a canal. Typically, the canal's maximum conveyance capacity will be associated with the normal depth, Y_{max}, for the maximum flow, Q_{max}. As described in section 2.7, flow changes will create depth fluctuations in the canal. When the depth is already at maximum, these fluctuations must be kept small to avoid overtopping. Therefore, flow changes must be gradual when the flow is at or near design capacity.

When canal flow is below channel capacity, normal flow could exist at a lower depth than the design depth of the channel. In the canal on figure 2-33, flow Q_i could be conveyed at normal depth, Y_i. Therefore, only the lower part of the channel is required for conveyance. The rest of the channel above the depth Y_i is essentially available as storage. Typically, canals will be operated at depths near the maximum level—even at low flows. Checking the water up to maximum depth, Y_{max}, takes advantage of the in-channel storage available. In-channel storage provides a buffer to absorb flow changes because flow imbalances can be accommodated by supplying or depleting this storage volume.

For reasons discussed above, acceptable rate of flow change for a canal will be a function of flow at the time of the change and size of the canal prism. At any given flow rate, the maximum rate of flow change could be established and operations restricted accordingly. New canals can be designed to include in-channel storage, by oversizing the conveyance channel beyond the minimum size required to convey maximum steady flow.

b. Off-channel storage.—Storage capacity outside the conveyance channel also will have a major impact on operations. Reservoirs, ponds, and tanks are commonly used to provide this storage. Storage facilities will provide a buffer to insulate the canal on one side of the storage facility from flow changes that occur on the other side. A mismatch or imbalance in flow can exist without wasting water

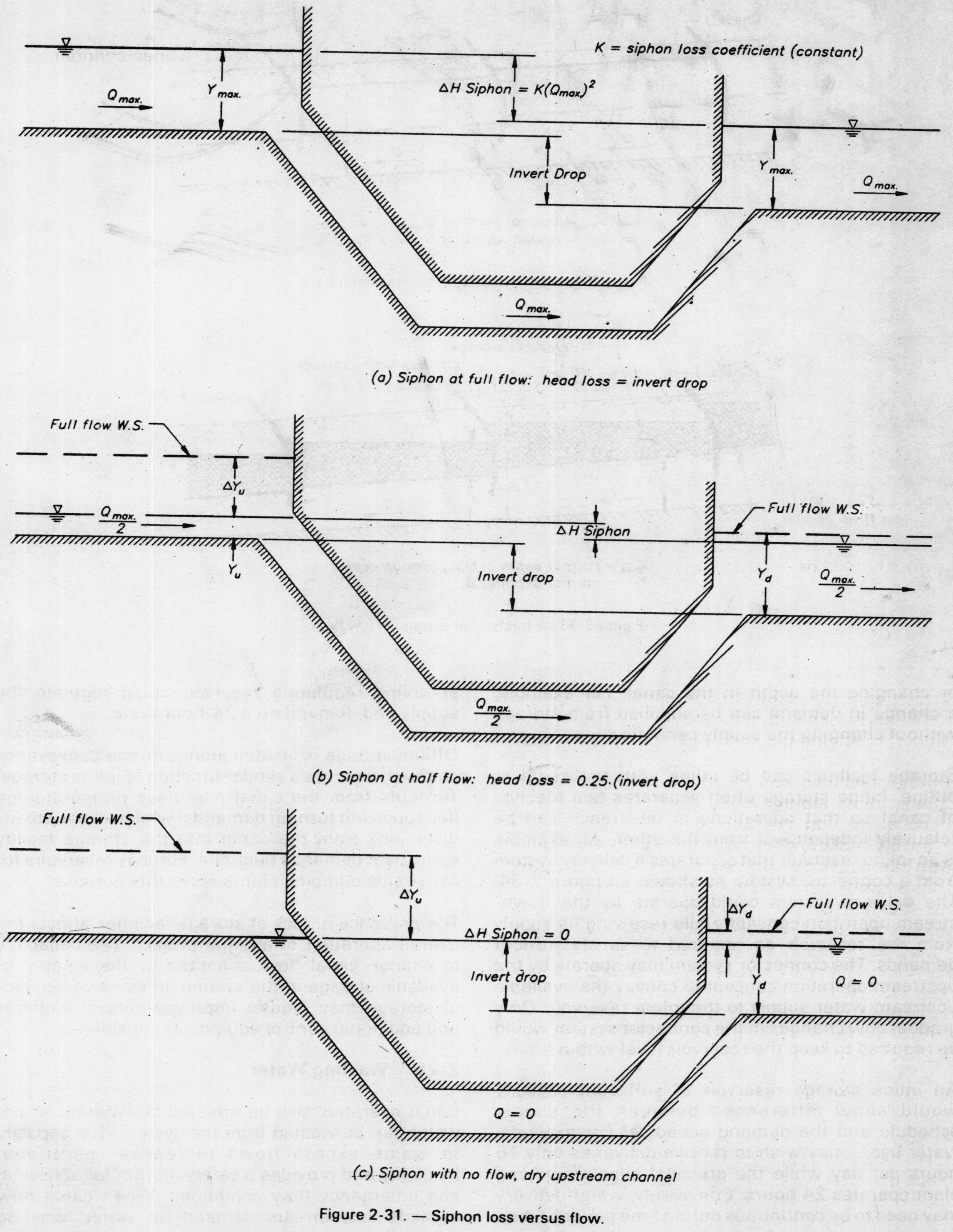

Figure 2-31. — Siphon loss versus flow.

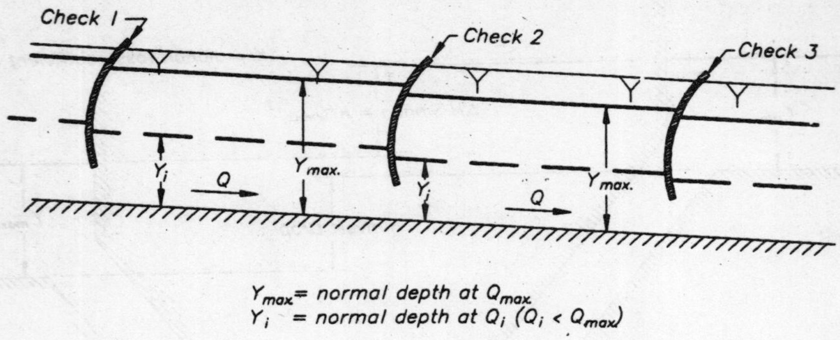

Figure 2-32. — Normal depth for maximum flow.

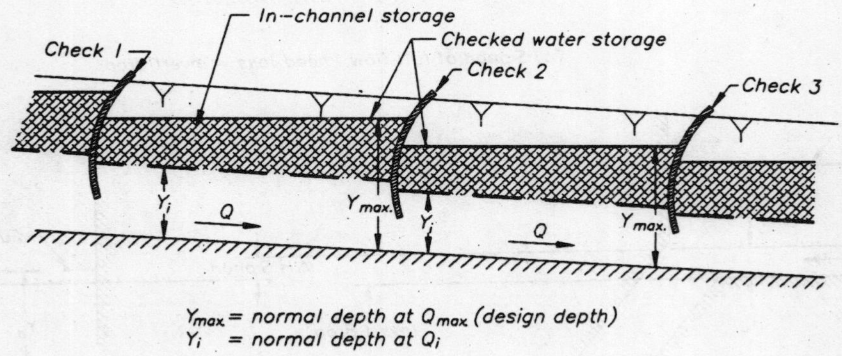

Figure 2-33. — In-channel storage at low flow.

or changing the depth in the canal. For example, a change in demand can be supplied from storage without changing the supply canal flow.

Storage facilities can be inline with the canal or offline. Inline storage often separates two reaches of canal so that operations in one reach can be relatively independent from the other. An example is an inline reservoir that separates a delivery system from a connector system as shown on figure 2-34. The delivery system could operate by the downstream operation concept while receiving its supply from the reservoir as required to satisfy turnout demands. The connector system may operate by the upstream operation concept to convey the available upstream water supply to the inline reservoir. Only gradual flow changes in the connector system would be required to keep the reservoir level within limits.

An inline storage reservoir of sufficient capacity would allow differences between the supply schedule and the demand schedule. For example, water users may wish to receive deliveries only 16 hours per day while the principal supply pumping plant operates 24 hours. Conversely, water delivery may need to be continuous during time periods when the supply pumping plant is shut down. In both cases, an inline regulating reservoir could regulate the supply and demand on a 24-hour cycle.

Offline storage is located alongside the conveyance channel; it serves a similar function to inline storage. Turnouts from the canal may have offline storage (for supplying turnout demands) which allows steady diversions from the canal into the storage facility while user demands fluctuate. Forebay reservoirs for canal-side pumping plants serve this purpose.

The presence or lack of storage facilities affects the overall operation of a canal project. The capability to change canal flow is limited by the amount of available storage in the system. In some cases, lack of storage may require improved control methods and additional control equipment capabilities.

2-25. Wasting Water

Canal operation will be affected by whether or not water can be wasted from the system. The capacity to waste excess flows increases operational flexibility and provides a safety margin for abnormal and emergency flow conditions. When canal flow exceeds downstream demand for water, wasting provides an outlet for the excess supply. Limits on

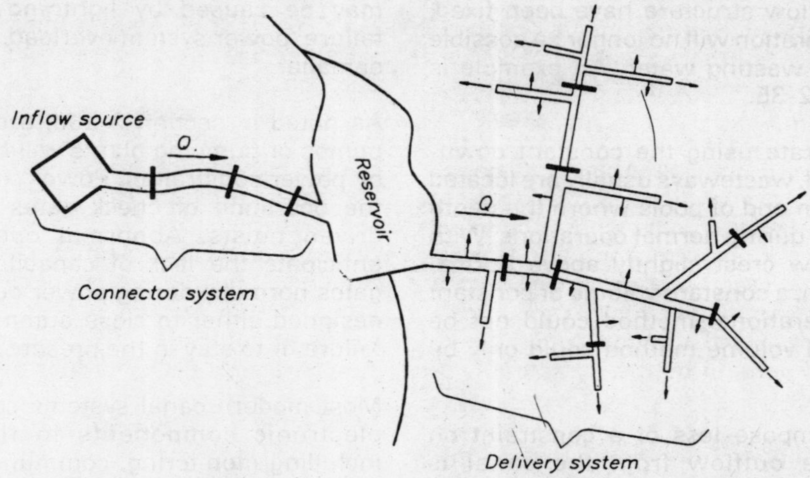

Figure 2-34. — Inline reservoir separating connector and delivery systems.

the amount of water wasted from a canal and wasteway location along the canal impose operational constraints.

Wasting is an undesirable loss of a valuable resource; it prevents water from being used for the intended purpose. In most canals, wasting water is unintentional during normal operations and is avoided whenever possible. However, by sacrificing some water, wasting can be a valuable operational tool during emergencies to prevent more detrimental consequences.

Usually, water is wasted either because flow into a canal has increased or because canal-side diversions have decreased. Storm runoff is a common cause for increased inflow entering a canal through drain inlets. This represents an uncontrolled situation where wasteways become valuable protective devices. Outflow decreases may stem from various normal or abnormal causes, such as turnout flow reduction, power failure, pump shutoff, gate(s) closing, or flow obstruction.

Many variables affect wasting of water from canals. Economics affect the amount and location of waste flows. The value of water will have a direct impact on whether to waste or store it. In a system that pumps water to a higher elevation, power costs are sacrificed whenever that water is wasted. The cost of constructing wasteways and of any damage caused by waste flows leaving the canal system also must be considered. In many canals, wasteways are designed for use only during emergency conditions. Wasteways often provide a more viable solution to protecting the canal system during an emergency than other more expensive alternatives.

Using wasteways can be practical for canals fed by gravity, especially those canals that can easily return waste flows to natural channels. Many canals are constructed nearly parallel to the river which supplies the canal, allowing excess flows to be returned to the river further downstream. These systems can be simple to operate by diverting a steady supply equal to the maximum demand and wasting the excess when demand decreases.

When excess flows cannot be wasted, they must be stored. The amount of excess flow then is limited by the amount of storage available. This creates a constraint on the amount of mismatch existing between inflow and outflow and the duration of this mismatch. A canal must be able to respond quickly enough to satisfy this constraint or hazardous conditions would follow. An interrelation will exist between capability to waste, storage capacity, and operational flow mismatches of a canal system.

Usually, wasteways are located so that excess canal flows can be passed into natural drainage channels to minimize any detrimental effects to the environment. The availability of suitable topography will affect waste from any given canal. Water quality considerations also may be a factor. In some instances, the canal water quality may totally preclude waste flows. An example is the Bureau's Garrison Diversion Unit in North Dakota where waste flows are not permitted because of environmental concerns for introducing water from one watershed into another.

Frequently, wasteways include a gravity overflow structure so water will be wasted if the canal level exceeds the level of the overflow. The level of the overflow crest is typically a small distance above the maximum normal operating water level so that water will be wasted automatically—whenever the water level gets too high. Once the crest elevation and location of the overflow structure have been fixed, some methods of operation will no longer be possible without undesirably wasting water. An example is illustrated on figure 2-35.

For canals that operate using the constant downstream depth method, wasteways usually are located near the downstream end of pools where the depth is relatively constant during normal operations. With a wasteway overflow crest slightly above normal depth, at this location, a constant volume or constant upstream depth operational method could not be used. The controlled volume method could only be used at low flows.

Gated wasteways impose less of a constraint on operations because outflow from the canal is controllable. Controlled wasting is less depth-dependent than with uncontrolled overflow wasteways. Location is still important, however, because water must be conveyed to the wasteway location before wasting.

Many types of wasteways exist; reference [11] has information on wasteways and wasteway design.

2-26. Power Restrictions

Power availability will have an effect on most canal systems. Canal system components that require electrical power include pumps, motorized gates, and monitoring, control, and communication equipment. All systems requiring electrical power are dependent upon power schedule restrictions and power outages.

Power restrictions may vary from occasional limitations on the amount of power used at any time to a total power loss. Some power restrictions exist as normal conditions, which are constantly present, such as limits on the maximum rate of power consumption. Another example of a normal power restriction is time-dependent availability; some canal systems cannot consume power during peak demand time periods. Abnormal power supply interruptions

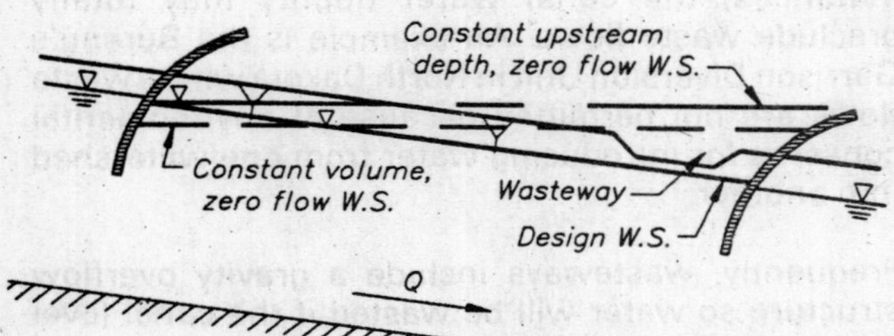

Figure 2-35. — Alternate methods of operation prohibited by wasteway at downstream end of pool.

may be caused by lightning strikes, equipment failure, power system overload, or transmission line damage.

As noted in section 2-23d, a canal system having pumps or pumping plants will be influenced greatly by power constraints. Power constraints also affect the operation of check gates with electric motor driven hoists. Abnormal operation plans must anticipate the lack of capability to operate check gates normally during power outages. Gates can be designed either to close automatically upon power failure or to stay in the present position.

Most modern canal systems contain electrical and electronic components in the control system, including monitoring, communications, and control equipment. Backup power often is provided so that a control system is not completely disabled during power failure.

In the construction of new canals and in modifying existing canals, power availability is a constraint dependent on economics. Constructing power lines in remote areas may be too expensive to warrant using electrical canal control equipment. Solar power is an alternative, but the limitations of solar power systems affect operational possibilities. For example, the number of gate movements per day may be restricted by the available power.

Operations must be tailored to conform to power restriction constraints. For a pumping plant having multiple pump units, the number of units operating at one time may be limited. The limit may vary according to time of day because the demand in the power network may affect the availability of peak power. Even without restrictions on the availability of peak power, the cost of power may be higher during peak demand times. Economics may justify scheduling operations to use less power during peak times and more during offpeak periods.

2-27. Personnel

The availablity of technical expertise is an important factor in canal operations. Competent operators are required who possess the technical qualifications and capabilities for successful operation. Past experience has proved that the most successful canal system operations are maintained by competent, skillful technical personnel.

Personnel requirements are different for the different methods of control. Local manual control requires a great deal of experience and judgement by the ditchriders in setting the main canal gates, adjusting turnouts, and in dealing directly with water users. Automatically controlled canals are less dependent on the ditchriders, but require skilled electronics technicians for equipment maintenance.

Staffing considerations may be paramount in determining the feasibility and success of alternative methods of operation and control. A method using existing personnel may be preferable to one requiring staffing changes. In some geographical regions, obtaining the level of technical expertise required to support automatic control equipment is not easily achieved. In other areas, automatic control may be preferable because of the lack of good ditchriders or high labor costs.

BIBLIOGRAPHY

2-28. Bibliography

[1] Burt, C. M., "Overview of Canal Control Concepts," *Planning, Operation, Rehabilitation, and Automation of Irrigation Water Delivery Systems*, ASCE, edited by D. D. Zimbelman, pp. 81-109, New York, 1987.

[2] Clemmens, A. J., "Delivery System Schedules and Required Capacities," *Planning, Operation, Rehabilitation, and Automation of Irrigation Water Delivery Systems*, ASCE, edited by D. D. Zimbelman, pp. 18-34, New York, 1987.

[3] Replogle, J. A., and J. L. Merriam, "Scheduling and Management of Irrigation Water Delivery Systems," *Irrigation Challenges of the 80's*, ASAE, pp. 112-126, Lincoln, Nebraska, 1980.

[4] Ploss, L. F., J. F. Buchheim, and L. A. Brower, "Irrigation Project Distribution System Scheduling," *Irrigation and Drainage in the 1980's*, ASCE, pp. 293-302, Albuquerque, New Mexico, 1979.

[5] Chow, V. T., *Open-Channel Hydraulics*, McGraw-Hill Book Co., New York, 1959.

[6] Chaudhry, M. H., *Applied Hydraulic Transients*, Van Nostrand Reinhold Company, New York, 1979.

[7] Wylie, E. B., and V. L. Streeter, *Fluid Transients*, McGraw-Hill International Book Company, New York, 1978.

[8] Clemmens, A. J., "Canal Capacities for Demand Under Surface Irrigation," ASCE, *Journal of Irrigation and Drainage Engineering*, vol. 112, No. 4, pp. 331-347, November 1986.

[9] Goussard, J., "NEYRTEC Automatic Equipment for Irrigation Canals," *Planning, Operation, Rehabilitation, and Automation of Irrigation Water Delivery Systems*, ASCE, edited by D. D. Zimbelman, pp. 121-132, New York, 1987.

[10] Clemmens, A. J., and J. A. Replogle, "Controlled-Leak Methods for Water Level Control," *Planning, Operation, Rehabilitation, and Automation of Irrigation Water Delivery Systems*, ASCE, edited by D. D. Zimbelman, pp. 133-141, New York, 1987.

[11] Aisenbrey, A. J., R. B. Hayes, H. J. Warren, D. L. Winsett, and R. B. Young, *Design of Small Canal Structures*, Bureau of Reclamation, rev. reprint, 435 pp., Denver, Colorado, 1978.

[12] Walker, R. E., "Long Crested Weirs," *Planning, Operation, Rehabilitation, and Automation of Irrigation Water Delivery Systems*, ASCE, edited by D. D. Zimbelman, pp. 110-120, New York, 1987.

[13] Buyalski, C. P., "Discharge Algorithms for Canal Radial Gates," Bureau of Reclamation Report No. REC-ERC-83-9, 232 pp., Denver, Colorado, December 1983.

[14] Rogers, D. C., "Operation Considerations in the Design of Canal Systems," *Design of Hydraulic Structures 89*, A. A. Balkema, pub.; edited by M. L. Albertson, and R. A. Kia, pp. 457-467, Rotterdam, 1989.

CHAPTER 3

CONTROL FUNDAMENTALS

CONTROL THEORY

3-1. Canal Control

The term *canal control* describes those steps necessary to ensure the required pool water level and flow along the canal. These conditions are controlled by adjusting the volume of water pumped into the canal, adjusting the positions of check gates, and by regulating flow through the delivery turnouts.

Canal conditions depend upon the adjustment of variables that effect control of the canal system. Some variables are easily adjusted while others are more complicated or cannot be adjusted. The variables easily adjusted are the gate positions, number of pumps supplying or diverting water, flow regulation of the pumps, and the diversion of water into and out of the canal system. Other variables such as weed growth, debris, syphons, and the geometry of the canal cannot be readily controlled. Canal control is achieved by manipulation of the variables to obtain the desired canal system conditions [1].

3-2. Control Classification

Control can be classified as either *open-loop* or *closed-loop*. Open-loop control is where the controlled quantity is adjusted with no comparisons to actual response or to desired conditions. Closed-loop control is where the controlled quantity is measured and compared with a reference or standard representing the desired performance. Any deviation from the reference is fed back into the control system so that it will reduce the deviation of the controlled quantity from the reference.

An example of open-loop control is a traffic light signal. The traffic lights respond at specific time intervals regardless of the actual traffic conditions. Closed-loop control is one where the traffic light senses the presence or flow of traffic and adjusts the time interval of the traffic lights according to traffic flow.

Feedback control is another term to describe closed-loop control. In canals, feedback type control systems are used to minimize the magnitude and duration of the mismatch between the supply and the demand.

3-3. Basic Elements

The elements of a simple feedback control system consist of the *sensor, comparator, control element* and *actuator*, as shown on figure 3-1.

The *sensor* provides input to the control system. The sensor also converts an observable parameter such as water level, flow, or gate position to a quantity that can be used by the control system. A canal control system may use—and often does—more than one sensor.

A *setpoint* is a reference input to the comparator element. A sensor input quantity that is different from the setpoint value causes the control element to initiate an output to the actuator. The setpoint reference value is sometimes referred to as the target value for canal control systems. The deviation from setpoint (target value) is called the error.

A *comparator* is a device that compares two inputs and provides an error (or difference) as an output. In a canal application, a comparator would compare the input value from the sensor (or sensors) with the setpoint value and produce the appropriate error signal to the control element.

The *control element* performs the same set of steps that a ditchrider does to adjust a canal system based upon observed conditions. The set of steps are defined by the rules or processes required for the solution to the problem. The control element performs these steps automatically and provides a signal to the actuator.

The control element often is referred to as the controller. Sometimes, the complete system, from the sensor to the actuator, is called the controller. This manual will use the term "controller" for the complete control system and "control element" for the logic control device or equipment that performs the control algorithm.

The *actuator* converts output of the control element to a mechanical operation that effects the process operation. In a canal system the actuator converts the output of the control element to an electrical, mechanical, or hydraulic action to: open or close check gates, open or close valves, and start or stop pumps.

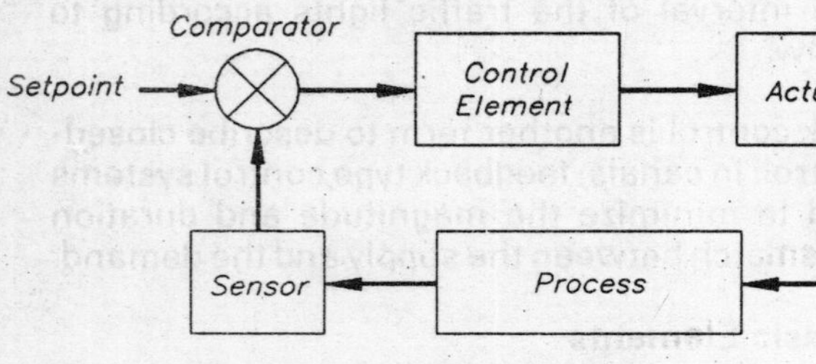

Figure 3-1. — Simple feedback control system.

3-4. Canal Controller Enhancements

The purpose of an automatic feedback control system is to reduce the error as rapidly and smoothly as possible. The simple feedback control system shown on figure 3-1 can be enhanced by adding more elements and changing the function of the elements within the feedback path as shown on figure 3-2. A wide variety of enhancements to the feedback loop functions are possible and many have been successfully applied to automatic canal control. Some enhancements include: water level filters, multiple sensor inputs, variable speed motor actuators, and more complex control algorithms. Oftentimes, additional manipulation of the data between the sensor and the actuator is required to obtain the desired canal operation.

Now that the basic elements of the feedback control system have been described and the major components of the primary control loop identified, certain control system enhancements will be described. The primary control element contains the set of rules or sets of instructions (algorithm) and calculates the desired value (position or speed) that the gate, valve, or pump must operate. A secondary control loop within the primary control loop (sensor, comparator and actuator) provides the actual control output value to position the gate (or valve) or sets the pump speed. For example, if a 2-inch (51 mm) increase in the gate opening is required, the actuator would turn the gate motor on for the time required to move the gate 2 inches in the appropriate direction and turn the gate motor off when the gate is at the desired position. The actuator produces the desired output of the control element and provides the input to the canal system.

The control algorithm will be designed to operate the canal system in an efficient, responsive, and

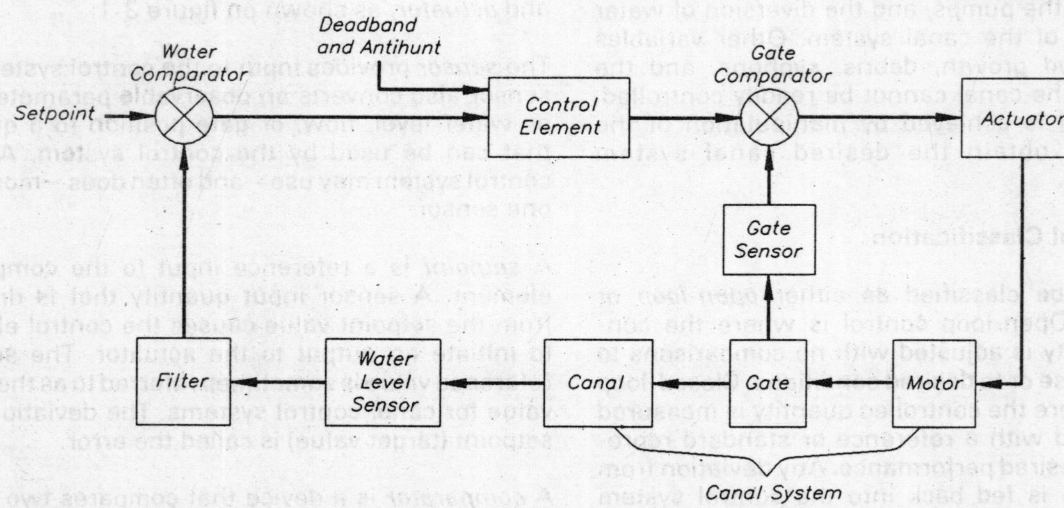

Figure 3-2. — Canal controller.

stable manner. For the algorithm to be designed to accomplish canal control in this manner, enhancements are contained in the control loop that specifically apply to canal system control. These functions (which follow) include: filters, deadbands, antihunt, and multicontrol loops.

a. **Filters.**—The input to a canal control feedback loop (the sensor output) sometimes is filtered before being processed by the controller. The primary purpose of the filter is to eliminate canal water surface wave action and also to decrease the number of gate movements. Additionally, filters can govern the stability of a control system by filtering out critical frequencies of the disturbances that tend to be amplified by the controller; this characteristic of filters is called *lag*.

Initially, water level filter elements were used which consisted of a small diameter tube that connected between the canal (or a stilling well) and a secondary well. Water level change in the secondary well lagged behind changes in the canal, because only a very small flow can pass through the small connecting tube.

Electronic or digital filters replaced hydraulic filters to achieve the same result. Instead of water surface elevation being filtered by the small diameter tube, electrical output of the water level sensor is filtered by a resistor-capacitor network or a digital computer program. The electronic or digital filter output then is used as the input to the canal controller.

b. **Deadband.**—A deadband is a predetermined range through which the measured signal can be varied without initiating a control action. It is the error allowed in the process to prevent the controlled device (gate) from continuously operating. Usually, deadbands are incorporated in the comparator element. The difference between the measured value and the setpoint reference value must be greater than the deadband before an output of the comparator will result.

Deadbands can be used with both the input from the sensor and the output to the actuator. When the input is a water level, a distance above and below the target value or desired water level will be prescribed as deadbands. If the water level input from the sensor (or filter) is within the deadband range selected, further control calculations or actions are not taken. On the output side of the feedback loop, a minimum gate movement may be established as a deadband. The calculated gate movement must be greater than the minimum before the actuator will operate the gate hoist motor. The deadbands also can be applied at other places within control algorithms. Sometimes, calculations performed by the controller are made dependent upon deadbands to minimize control actions (gate movements).

c. **Antihunt.**—Antihunt is an element used to define the part of the control process of disabling the actuator element when canal water level stops moving away from the setpoint and begins to return to the setpoint. This control action is used in some controllers to prevent excessive gate travel and instability. Antihunt improves stability by reducing the possibility of amplifying the disturbance. When using antihunt, a possibility exists that the water level may return only slightly toward the setpoint and then stop moving.

d. **Multicontrol loops.**—A canal control system is not restricted to a single control loop. By adding more sensors to the feedback control system shown on figure 3-1, a dual loop feedback control system is created. Figure 3-2 shows a canal controller having two control loops—a water level control loop and a gate position control loop. The gate position control loop is nested within the water level control loop. The control element output is the calculated gate position. The gate control loop automatically moves the gate to the calculated position.

3-5. Control Stability

Stability is a requirement of a control system for proper operation. A control system is considered to be stable if it returns to equilibrium after a change in the system occurs. An unstable system causes oscillations to amplify or persist instead of attenuating. Oscillations can result from a resonance condition where the natural frequency of the hydraulic system is the same as the frequency of control actions. Instability also can be caused by overcorrection when the magnitude of corrective actions is too large for the amount of error.

An ideal control system should rapidly correct errors and then return to the desired flow condition. This requires both *responsiveness* and *stability*. A very responsive system will correct errors quickly, but it also may be unstable. A compromise must be reached to optimize these two requirements.

A slow and stable control response may work best in a canal that generally experiences minor or gradual flow changes. This response will prevent unnecessary gate movements, reduce the frequency and duration of movements, and avoid overcorrection. The control system can be relatively sluggish without falling too far behind because water depths will not change too fast.

Conversely, a canal that can experience rapid, large flow change requires a more responsive control system. Corrective action must be taken before the water levels deviate too far from the setpoint. A control system must be adjusted in the best possible manner for efficient canal operation by adjusting parameters so that it is completely responsive to severe demands, but still stable during the normal day-to-day demand fluctuations.

BASIC CHARACTERISTICS OF AUTOMATIC CONTROL

3-6. Modes of Control

Automatic controllers for canal systems operate to restore some controlled variable to a desired value. The manner in which they operate is called the *mode of control*. The most common modes of control are:

1. Two Position
2. Three Position
3. Proportional
4. Reset (integral)
5. Rate (derivative)

The above modes of control are often combined to obtain the desired control action. A common combined mode controller is the PID controller. The PID controller combines the Proportional, Integral (reset), and Derivative (rate) modes to perform the desired control action. Each mode of control listed above has been used by the Bureau of Reclamation to control canals. All automatic controllers will operate according to one or more of the above modes. A more detailed discussion of each mode of control follows.

a. Two position.—Two position control responds to the deviation from the setpoint by operating the actuator to either of two extreme positions. The controller will not move the actuator to any intermediate position between the two extremes. Because the two positions are fully on and fully off, the control is referred to as On-Off control. The sensor provides the input information to the comparator and the deviation from the setpoint or error output from the comparator provides the actuator with only two outputs representing the two extreme positions of the controlled device.

Figure 3-3 illustrates the action of a two-position controller. In this example, the process of maintaining a constant water level in a tank is illustrated. Water is withdrawn from the tank on demand. When the level in the tank falls below the desired or setpoint level, a valve is opened fully to fill the tank. The valve is fully closed when the desired water level in the tank is reached. A *deadband* would have to be selected about the setpoint value for the two-position control to operate without excessive cycling.

Two-position or On-Off control is not practical for canal check structure operation. It is applicable to canal pumping plants, pipe distribution systems, sump pump operation, and solenoid operated valves.

b. Three position.—Three position control responds to the deviation from the setpoint (error) by operating the actuator for a predetermined amount of time. Three position control also is referred to as *floating* control and *set-operate-time/set-rest-time* control. Figure 3-4 illustrates a system that maintains the water level in the tank between the two float sensor contactss.

Three position control results from the fact that the final control element may exist in one of three states. These states are:

1. Off—no corrective action
2. On—above the setpoint—move final control element in the open direction just long enough to reestablish setpoint
3. On—below setpoint—move final control

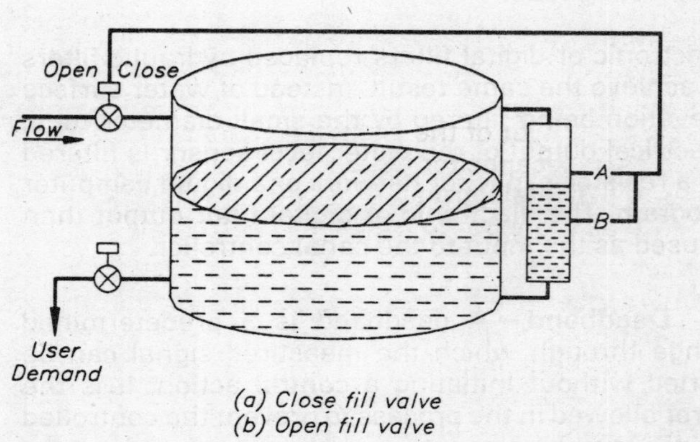

Figure 3-3. — Two position control.

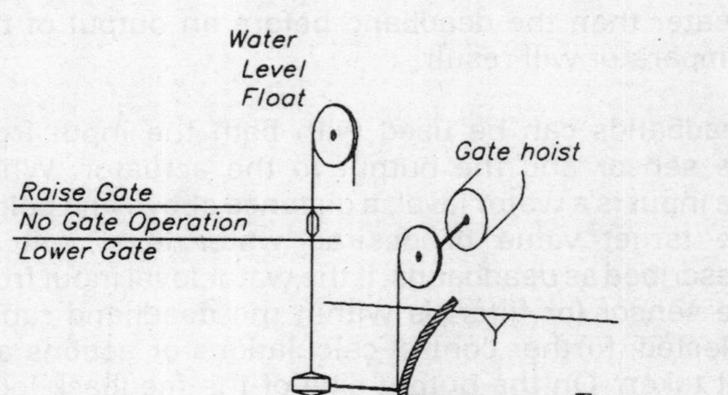

Figure 3-4. — Three position control.

element in the close direction just long enough to reestablish setpoint

The final control element does not operate as long as the controlled variable is within the deadband. Usually, the final control element operates at a much slower rate than for two-position control because intermediate positioning is desired. The corrective action must be set fast enough so that the system responds to the error and restores the setpoint, but slowly enough to prevent the controller from overcompensating and becoming an On-Off controller. The speed at which the controlling element moves is critical for a process using the three position control mode.

Controlled canal water surfaces have a tendency to cycle; they can be minimized and often eliminated by the width of the deadband and by selecting the proper speed for corrective action. Significant delays in system response, or rapid supply or demand changes will tend to cause water level cycling to become worse. Therefore, three position control is most appropriate for systems not having appreciable delay in system response and where system supply and/or demand can be satisfied with gradual changes in the operating conditions.

To achieve a match between the *input* (supply) and the *output* (demand) using three position control is extremely difficult. The three position control mode does not relate the values of the output to the input but senses the direction of the controlled variable. Matching the input to the output or matching the supply to the demand in a canal system is nearly impossible to achieve with the discontinuous corrective action performed by this mode of control. Therefore, three position control has limited application on canal systems with complex demand schedules.

c. **Proportional.**—Proportional control responds to the deviation from the setpoint using a fixed linear relationship between the value of the controlled variable and the position of the final controlled device. The proportional controller moves the final controlled device to a definite position for each value of the controlled variable. The amount of deviation from the setpoint represents the error and this error represents the magnitude of the output. A definite relationship can be established to couple the input to the output of the process.

An example of a proportional controller is shown on figure 3-5. The flyball governor is a proportional controller that causes the throttle valve to operate proportionally to the rotational speed (r/min) of the motor shaft. When the speed changes, the flyballs change position and cause the linkage to the throttle

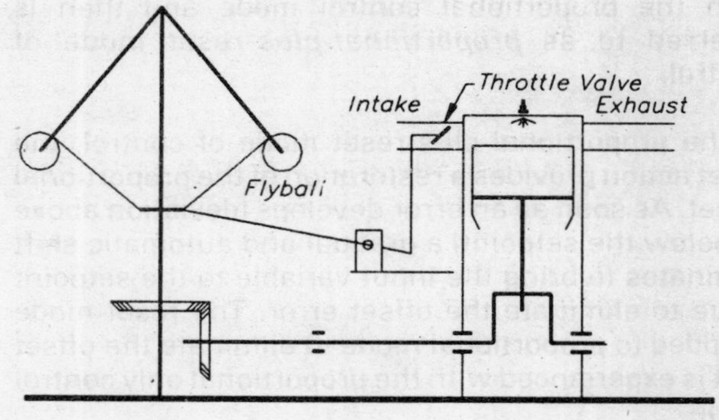

Figure 3-5. — Example of proportional control.

to be adjusted and thereby maintain a constant motor speed. This is reverse proportional control in that an increase in r/min causes the throttle to close.

In canal control, proportional mode of control operates the check structure gate based upon the amount of deviation (error) between the measured water level and the setpoint (target value) and a proportionality constant. The final control action from the actuator is to move the controlled device in direct proportion to the error amount. A residual error, characteristic of proportional control, remains and is referred to as water level *offset* in canal control applications. The proportionality factor is sometimes referred to as the *gain* of the control system.

Most proportional controllers have an adjustable gain. The gain determines the amount of gate movement for any change in the canal water level. Therefore, a rule of thumb: the larger the gain, the greater the change in gate position for any given change in the water level.

The term "proportional band" is sometimes used in reference to proportional control. Proportional band is the reciprocal of gain: the larger the gain the smaller the proportional band.

The major disadvantage of the proportional control mode is that a system demand change results in an offset. An offset is a sustained deviation of the water level from the setpoint as a result of the demand change. The magnitude of this offset may present difficulties for canal systems having gravity type turnouts.

d. **Reset or integral.**—Reset control responds to a deviation from the setpoint by collecting a time history of the error or integrating the error. The magnitude of the reset control action depends on the accumulation of corrective action and is based upon the magnitude and duration of the error. Reset control seldom is used alone. Usually, it is combined with the proportional control mode and then is referred to as *proportional-plus-reset* mode of control.

In the proportional-plus-reset mode of control, the reset action provides a restoration of the proportional offset. As soon as an error develops (deviation above or below the setpoint) a gradual and automatic shift originates to bring the input variable to the setpoint value to eliminate the offset error. The reset mode is added to proportional mode to eliminate the offset that is experienced with the proportional only control mode.

e. **Rate or derivative.**—Rate control responds to time rate of change in the magnitude and direction of the error deviation. Note that rate control has no effect on the process when the rate of change of the error is zero—even if the actual error is not zero. Additionally, rate action can cause a control response even when the actual error is zero. Therefore, rate control mode is not used alone, but in combination with proportional, proportional-plus-reset, or another control mode.

Rate control provides an initially large response when the deviation from setpoint occurs, so that the final controlled device is operated longer at first than it normally would without rate control. Then, having made this large initial change, the other modes of control take action to remove this effect and position the final control element as a proportional or proportional-plus-reset control action. Rate control provides a temporary overcorrection to the system that is proportional to amount of the rate of deviation per unit time.

Most often, rate control is used in process control systems that have an unusually large delay or lag time. Canal systems have this characteristic. The proportional gain must be set exceptionally small, and the reset is unusually slow. The addition of rate control to canal control algorithms should prove beneficial to canal control. However, because of complexity in determining the control parameters, additional research in the application of rate control is required.

In summary, several modes of control are applicable to canal control. Mode of control selection is based upon canal operational characteristics and required operational objectives. The simplest mode of control that will obtain the desired results is the best one to use. The modes of control that are applicable to canal control usually are selected based upon the method of operation. The three position control mode is applied most widely to canals that use the constant upstream depth method of operation. Proportional-plus-reset mode of control is suitable for all methods of operation.

3-7. Direct and Reverse Action

Automatic controllers can perform both direct action or reverse action control operation. The type of control action must be selected to be compatible with the upstream or downstream control concepts.

Direct action control is used for the upstream control concept. The water level inputs are located upstream of the controlled gate. When the water level or volume increases indicating a positive error, the check gate must open to release more water through the check structure and lower the upstream water level. In direct action control a positive error produces a positive change in output. The gain parameters for the controller are positive. Figure 3-6 shows an example of upstream direct action control.

Reverse action control is used for the downstream control concept. Water level inputs are located downstream of the controlled gate. When the water level rises or volume increases, indicating a positive error, the check gate must close to reduce the water flow into the canal segment. In reverse action control, a positive error produces a negative change in output. The gain parameters for the controller are negative. Figure 3-7 shows an example of downstream reverse action control.

3-8. Algorithms

An algorithm is a prescribed set of well defined rules or processes for the solution of a problem in a finite number of steps. In the control of canals, algorithms are the procedures (e.g., Little-Man, EL-FLO, and P+PR) used to implement the various canal methods of operation. The algorithm is designed to process the input information from the *sensors*, perform the *comparator* function, and calculate the proper output to the *actuator*. The input is quantities that are observed, measured, or predicted; the output is a control action.

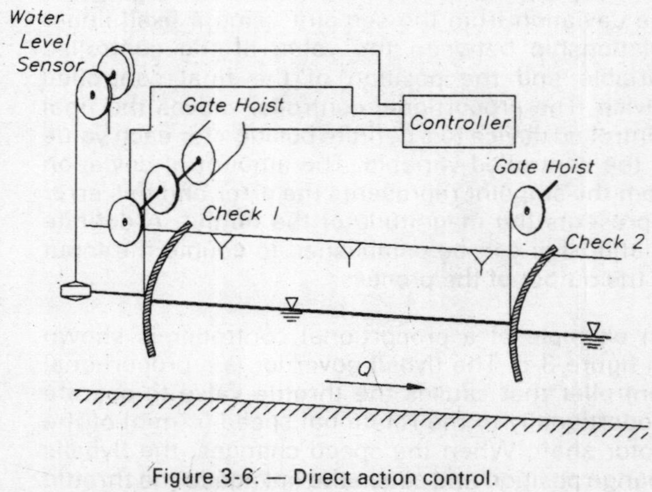

Figure 3-6. — Direct action control.

In the case of a ditchrider controlling a conventional canal, input is the observed water depth upstream of a check structure and position of gates in the check structure(s). By comparing the actual canal depth to the desired depth and flow schedule, the ditchrider can determine the new gate position. If the upstream canal water level is too high, the gate should be raised. The raised amount will be a function of the water level error, present gate opening, and perhaps other factors. The output of this reasoning process is a new gate position which hopefully will achieve the desired result of correcting the water level offset.

Thus, an algorithm is based upon the reasoning process of the ditchrider and can be expressed in the form of logic and mathematical equations. Many different techniques of canal control have been developed and have been expressed using algorithms. Algorithms that have been used successfully in canal system applications are discussed later in this chapter.

CANAL SYSTEM CONTROL METHODS

The preceding sections of this chapter have discussed basic control theory and control modes. Application of automatic control to canal systems requires an evaluation of the various methods for canal system control. For one to apply the correct control mode to a canal system, various methods for implementation must be understood. This section will expand on the canal control methods introduced in chapter 1 so that the appropriate canal control method can be implemented properly on an actual canal system. As introduced in chapter 1, four basic methods describe how canal systems can be controlled; they are discussed below.

3-9. Local Manual Control

Local manual control is the most prevalent control method used in conjunction with conventional operations. The canal operators, or ditchriders, travel up and down the canal adjusting check gates and turnouts en route. Labor saving machines, such as motorized gate hoists (with push button control stations), often are used to assist the ditchrider. Still, the control is performed manually and separately at each control structure. Local manual control is illustrated on figure 3-8.

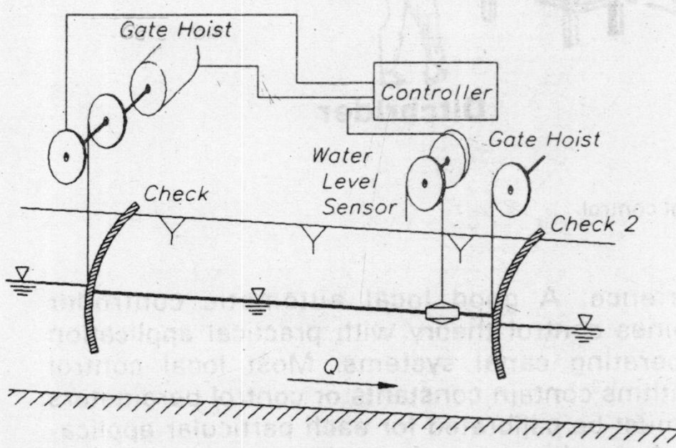

Figure 3-7. — Reverse action control.

On a large canal system, many ditchriders may be required. Usually, the canal system is divided so that each ditchrider is responsible for a portion of the total system. Individual ditchriders have to maintain the proper adjustment of all canal check gates and turnouts within their area. An experienced ditchrider can anticipate the arrival of flow changes at each of the structures which they control and time the adjustments accordingly.

Also, the ditchriders can exchange canal data, such as water levels and gate positions, by communicating on mobile radio to the watermaster at a central headquarters and to other ditchriders. This allows coordination of operations among the team of watermasters and ditchriders in the field. The quality of canal operations depends on the skill of the team operating within the limitations of the canal system.

3-10. Local Automatic Control

The local automatic control method is accomplished with automatic control equipment located onsite, as depicted on figure 3-9. Various types of control equipment can be used, from simple float-driven devices to complex arrangements of mechanical, electrical, electronic, and microcomputer components. Normally, local automatic control is accomplished without operator intervention.

A local controller senses one or more items of information such as water depth or gate position, and uses a well defined set of rules to produce a control response. Sometimes, this control logic can be built into a mechanical device, such as a float, switches, and counterweight attached to a gate. More often, this logic will be expressed as a mathematical equation called a *control algorithm*. As discussed in section 3-8, algorithms interpret the sensed information and calculate a desired control action such as a gate movement. A local automatic controller consists of all the equipment required to execute the control algorithm and the necessary inputs and outputs in a stand alone unit—usually located at each check structure.

Many different algorithms have been used in local automatic canal controllers. Most of them are used to adjust a check gate, based upon the water level measured at a single point in the canal, in an attempt

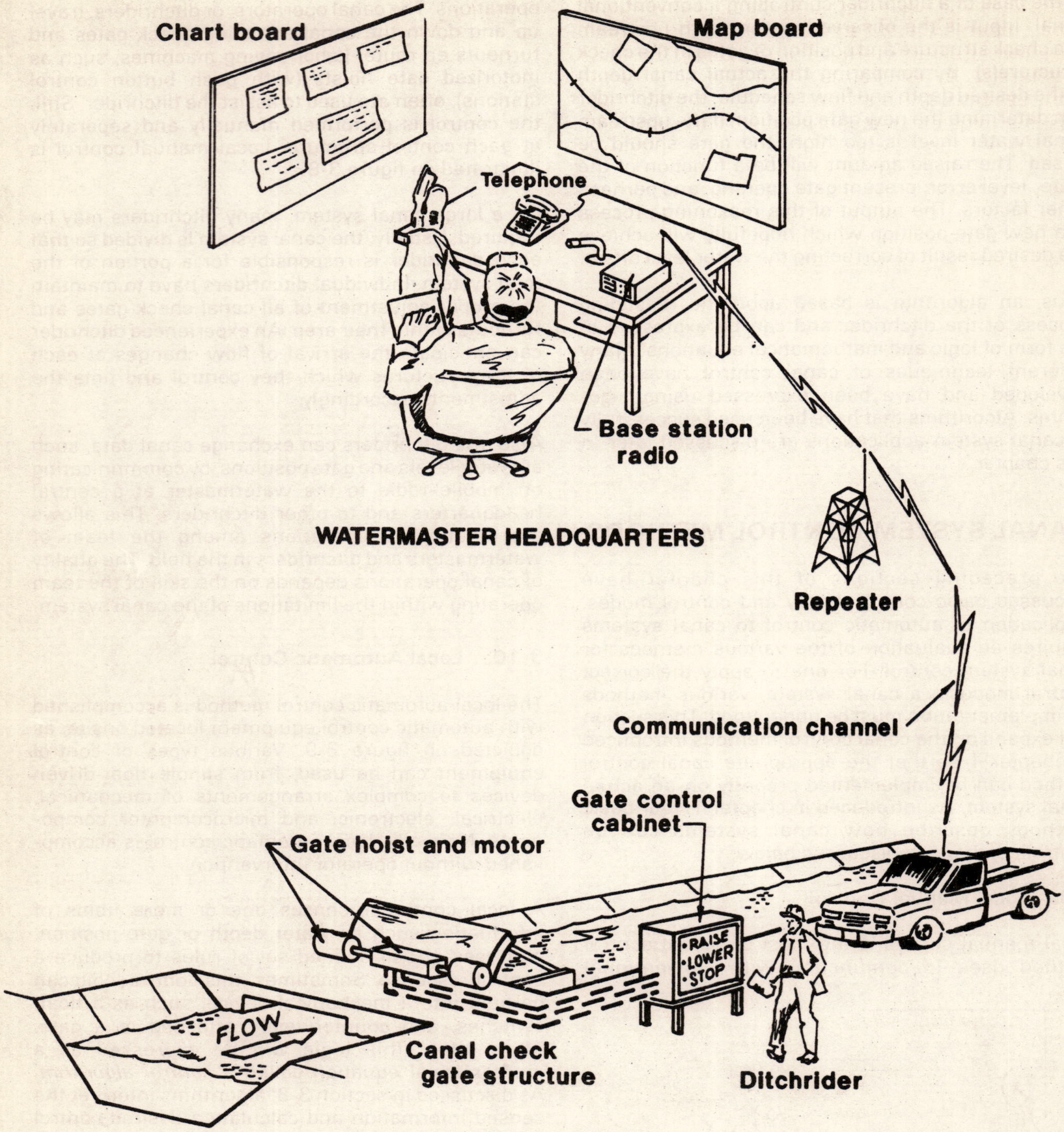

Figure 3-8. — Local manual control.

to maintain a constant depth at that point where the water level is measured. References [2, 3] describe two examples of automatic controllers using local control algorithms.

The design of a local automatic control system requires both technical knowledge and practical experience. A good local automatic controller combines control theory with practical application to operating canal systems. Most local control algorithms contain constants or control parameters that must be calibrated for each particular application. A combination of control engineering and operator experience is needed to design, build,

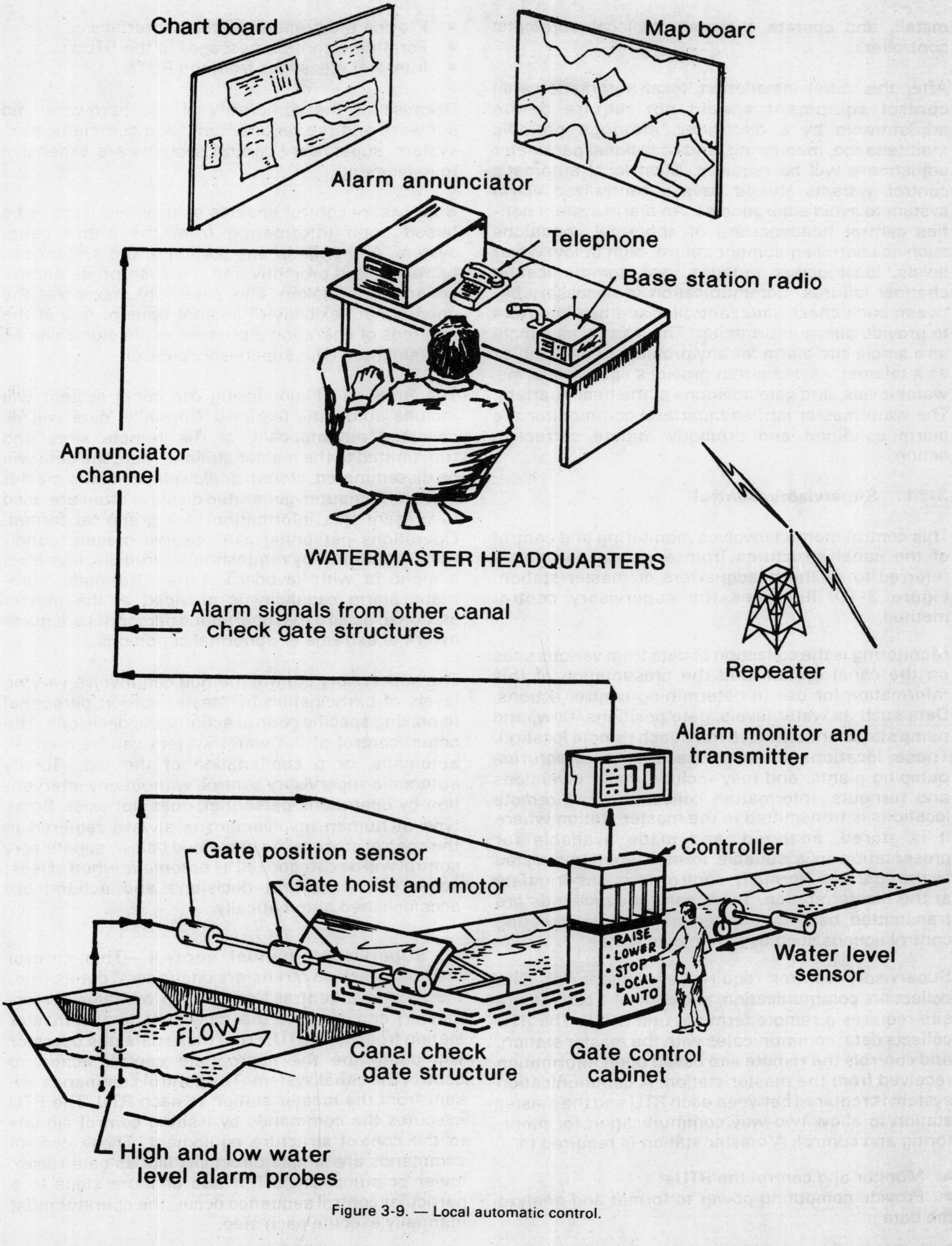

Figure 3-9. — Local automatic control.

install, and operate a system of local automatic controllers.

After the initial installation, local automatic canal control equipment should not require onsite adjustments by a ditchrider, although periodic maintenance, monitoring, and occasional parameter adjustment will be required. Most local automatic control systems should have a centralized alarm system to avoid emergencies. An alarm system notifies central headquarters of abnormal conditions such as control equipment failure, high or low water levels, local power outages, and communication channel failures. Communication is necessary between each check gate controller and headquarters to provide alarm information. This can be as simple as a single site alarm for any problem or as complex as a telemetry system that provides specific alarms, water levels, and gate positions at the headquarters. The watermaster (at headquarters) can monitor the alarm condition and promptly initiate corrective action.

3-11. Supervisory Control

This control method involves monitoring and control of the canal structures from a central location—referred to as the headquarters or master station. Figure 3-10 illustrates the supervisory control method.

Monitoring is the collection of data from various sites on the canal system and the presentation of this information for use in determining control actions. Data such as water levels, gate positions, flow, and pump status are collected from each remote location. These locations include check gate structures, pumping plants, and may include major diversions and turnouts. Information collected at all remote locations is transmitted to the master station where it is stored, analyzed, and made available for presentation in a suitable format when requested by the operator. Normally, control decisions originate at the master station. Then, control commands are transmitted back to the remote sites by creating control actions such as gate movements.

Supervisory control requires equipment for data collection, communication, and control. Each remote site requires a remote terminal unit (RTU). The RTU collects data, communicates with the master station, and controls the remote site based upon information received from the master station. A communication system is required between each RTU and the master station to allow two-way communication for monitoring and control. A master station is required to:

- Monitor and control the RTUs
- Provide computing power to format and analyze the data
- Provide the human/machine interface
- Formulate control messages to the RTUs
- Interpret messages from the RUTs

Because of the complexity of the hardware and software and the requirement for a communication system, supervisory control systems are expensive to implement.

Supervisory control enables control decisions to be based upon information from the entire canal system. A change in any portion of the system can be recognized promptly and the appropriate control action can be taken. This capability maximizes the operational flexibility of a canal system. Any of the methods of operation discussed in chapter 2 can be implemented with supervisory control.

The function of monitoring the canal system will include automatic features. Normally, data will be collected automatically at the remote sites and transmitted to the master station. Then, the data will be disseminated, stored, and displayed at the master station. Computer-generated displays often are used to present this information in a graphical format. Operations personnel can observe present conditions and trends by requesting desired displays from a menu or with keyboard entries. Normally, automatic alarm capability is provided at the master station to assure that operations personnel are made aware of extreme or abnormal conditions.

The supervisory control method can involve varying levels of participation by master station personnel in making specific control actions and decisions. The actual control of the water system can be manual, automatic, or a combination of the two. Totally automatic supervisory control, without any intervention by operations personnel, does not exist. Some type of human involvement is always required in the control process. As explained below, supervisory control will be categorized as *automatic* when at least some of the control decisions and actions are accomplished automatically.

a. Supervisory manual control.—This control method effectively transfers ditchriders' duties from the field to a central location. An operator (watermaster) can monitor the data sent to the master station from each RTU. From these data, the operator will determine the appropriate control action to control the canal system. The control commands are sent from the master station to each RTU. The RTU executes the commands by issuing control signals to the control structure equipment. These control commands are simple directives like as gate raise/lower or pump On/Off. If two or more steps to a particular control sequence occur, the operator must manually execute each step.

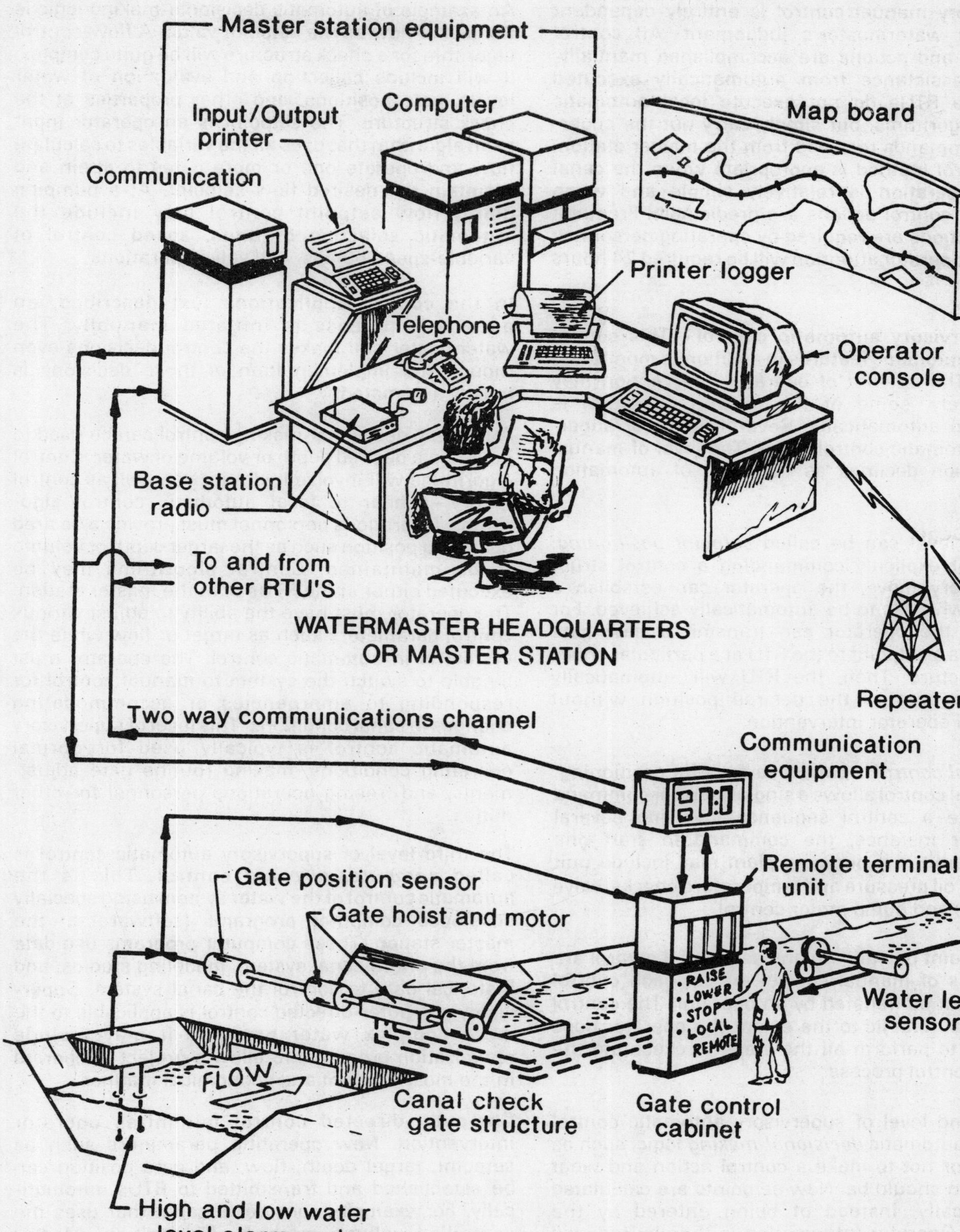

Figure 3-10. — Supervisory control.

Supervisory manual control is entirely dependent upon the watermaster's judgement. All control decisions and actions are accomplished manually, without assistance from automatically executed logic. The RTUs do not execute local automatic control algorithms, but simply carry out the supervisory commands received from the master station. This control method is appropriate when the canal system operation is relatively simple and when results of control actions are predictable. Frequent control actions are required by operating personnel; typically, operator attention will be required 24 hours a day.

b. Supervisory automatic control.—This control involves automatic features in addition to monitoring of data. The amount of operator intervention may vary widely; some of the routine operation is performed automatically. Several levels of supervisory automatic control occur. The level of manual participation declines as the level of automation increases.

The first level can be called *setpoint positioning*. Instead of explicitly commanding a control structure's every move, the operator can establish a setpoint which is to be automatically achieved. For example, the operator can transmit a new gate position as a setpoint to the RTU at a particular check gate structure. Then, the RTU will automatically move the gate to the desired position without additional operator intervention.

Sequential control is similar to setpoint positioning. Sequential control allows a single operator command to execute a control sequence involving several steps. For instance, the command to start one pumping unit in a pumping plant may include unit selection, oil pressure and temperature checks, valve operation, and pump motor control.

Both setpoint positioning and sequential control are extensions of manual control in that each control action must be initiated by an operator. The control action is automatic to the extent the operator does not have to perform all the steps in executing the desired control process.

The second level of supervisory automatic control includes automatic *decisional-making* logic, such as whether or not to make a control action and what that action should be. New setpoints are calculated automatically, instead of being entered by the operator. Operator intervention is thereby reduced even further. At this level, the operator will periodically adjust the automatic control and may switch to setpoint or manual control during emergency.

An example of automatic decisional-making logic is the use of flow as the setpoint value. A flow-control algorithm for a check structure will be quite complex. It will include collection and evaluation of water levels, gate positions, and other properties at the check structure. The setpoint is an operator input to an algorithm that uses all the variables to calculate flow and operate one or more gates to attain and maintain the desired flow setpoint. At a pumping plant, flow setpoint control may include the automatic selection of units, speed control of variable-speed pumps, and valve operations

In the control applications just described, an automatic process is initiated manually. The watermaster still makes the control decisions even though the implementation of those decisions is mostly automated.

Automatic decisional-making control can be used to maintain a desired depth or volume of water. Control algorithms will involve hydraulics as well as control theory—similar to local automatic control algorithms. Operations personnel must provide a desired operating position such as the target depth or volume to be maintained. Control algorithms may be executed either at the RTUs or at the master station. The operator must have the ability to adjust various control parameters such as target or flow while the system is in automatic control. The operator must be able to switch the system to manual control for responding to emergencies or accommodating changes in canal conditions. This level of supervisory automatic control is typically used for normal operating conditions, making routine gate adjustments, and freeing operations personnel for other duties.

The third level of supervisory automatic control is called *computer-directed* control. This is the automatic control of the water system using specially developed computer programs (software) at the master station. These computer programs use data from the entire canal system, modeling studies, and historical data to control the canal system. Supervisory computer-directed control is applicable to the more complex water projects; it may include optimization programs to allow a project to operate in the most efficient and economical manner.

Computer-directed control minimizes operator intervention. New operating parameters such as setpoint, target depth, flow, and gate position can be established and transmitted to RTUs automatically. For example, consider a canal that uses the controlled volume method of operation. Master station software could automatically calculate new target volumes for all the canal pools based upon information such as water and power supply/demand schedules, system operational criteria, and past and present depths and flows. These target volumes then could provide the new operating parameters for the automatic decisional-making control described above.

3-12. Combined Control Methods

Many canal systems use a combination of two or perhaps all three of the above control methods. The control method can change as the operator desires—with time or with location. The most common combination is the use of manual control as a backup for automatic control when power loss, equipment failure, or other emergency situations occur. Any automatically controlled system, no matter how sophisticated, should disable and revert to manual control when control logic fails or an error in equipment operation is detected. Any change in the status of automatic operation should be brought to the attention of the watermaster by an alarm.

Manual and automatic control are sometimes used simultaneously at different locations on a canal system. Canals that are predominantly manually controlled may have local automatic controllers in particular locations. For example, a gravity turnout may require a local automatic controller to maintain a constant delivery rate if the main canal water level changes frequently. A ditchrider may not be able to visit the site often enough to make the necessary gate opening adjustments. When turnout flow changes are desired, the ditchrider would enter a new flow setpoint into the local controller instead of manually adjusting the turnout gate. Hence, this system combines the local manual and local automatic control methods.

Combining local automatic control with supervisory control often will have merit. By using RTUs with stand-alone control capability, the frequency of communication with the master station can be greatly reduced. This can save master station computers and/or operators from constantly sending control information to each check structure on the canal. With this combination, supervisory control primarily may be used to adjust setpoints or control parameters in the local control algorithms. Supervisory control also can override the local control whenever abnormal conditions arise. The system could operate similarly to an automobile cruise control, wherein the local controllers "cruise" except when overridden by the master station. This combination provides an additional benefit for allowing normal local automatic control to proceed without interruption during a temporary communication failure between RTU and master station.

CONTROL ALGORITHMS

The Bureau of Reclamation has developed several algorithms for application to the automation of canals. Addtional algorithms have been developed by others. The following canal control algorithms have been used in canal system automation projects constructed by the Bureau and others. In the following, each control algorithm will be discussed in terms of the method of operation and the control mode for which it was designed. The algorithms that will be discussed are:

1. Little-Man
2. Colvin
3. EL-FLO and EL-FLO + RESET
4. P+PR
5. BIVAL
6. Dynamic Regulation
7. Gate Stroking
8. Variable Target

3-13. Little-Man Algorithm

The *Little-Man algorithm*, based upon the three position (floating, set-operate-time/set-rest-time) control mode, has been used extensively for water level regulation of many canal systems. The electromechanical type Little-Man controller has been the most predominate but also can be implemented using a microprocessor to execute the programmed algorithm.

In reference to figure 3-11 (the electromechanical type of Little-Man controller), usually the sensor consists of a float operated trip bar or cam. The comparator unit consists of a simple arrangement of microswitches. One microswitch is set to operate at the upper limit and the other at the lower limit of the deadband. Usually, the setpoint is the design depth of the canal for the maximum flow conditions. The deadband limits are typically 1/4 inch (6.1 mm) from the setpoint. The control element consists of two timers that are activated by the comparator unit raise and lower microswitches. The timers energize the raise or lower relays to the motor actuator for only a few seconds and then an interval of rest time is provided between gate movements.

The operation of the control element is referred to as *set-operate-time/set-rest-time* (SOT/SRT). The adjustable timers provide the flexibility to select the run/rest intervals to match the anticipated disturbances. However, in canal systems, disturbances are usually variable. Therefore, to achieve a match between the supply and demand of the canal system on a continuous basis, the run/rest timing sequence would have to be adjusted continuously to prevent the system from becoming unstable. Figures 3-11 and 3-12 show schematic diagrams for two types of Little-Man controllers.

The *antihunt* device (described in sec. 3-4c) allows the Little-Man controller to be used for a wider range

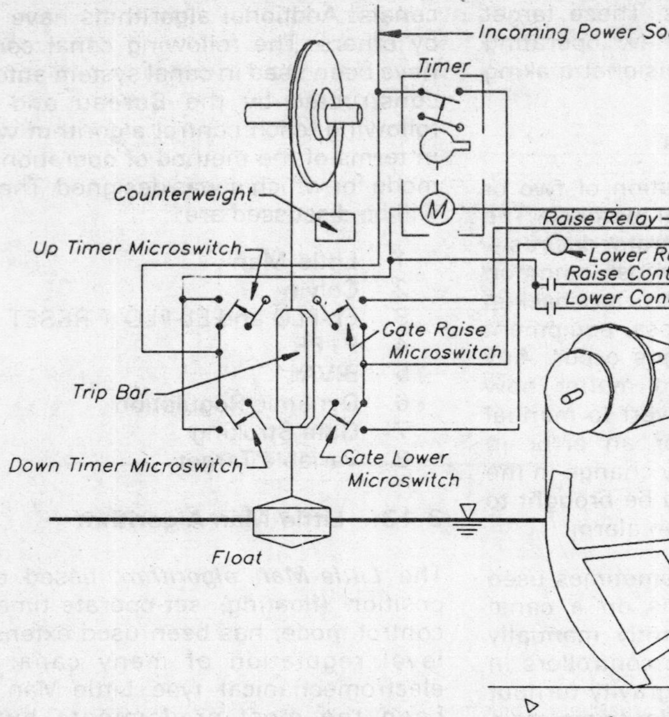

Figure 3-11. — Microswitch operated Little-Man controller.

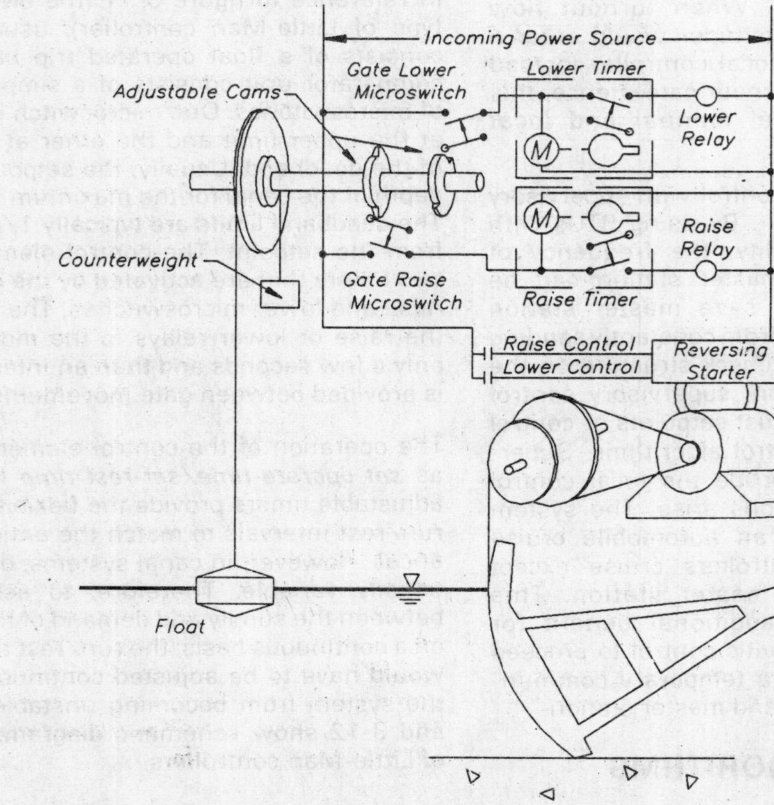

Figure 3-12. — Cam operated Little-Man controller.

of disturbances using a single speed SOT/SRT output. The antihunt device prevents excessive gate travel and reduces the amplifying effect on the disturbance, thus providing a more stable control system.

The operation of the Little-Man controller can be expanded by the addition of multiple sets of run and rest timers. These additional timers can be operated for larger and smaller water level deviations from the setpoint relating the magnitude of the disturbance to rate of change in the gate movement. Deadbands are used to select each different set of run/rest timers. As the water level deviates from the setpoint beyond the upper or lower deadband limits, the multistage timers cause the gate movement time to increase and the rest time interval to decrease. Once the water level begins to return toward the setpoint, the antihunt mechanism deactivates further gate movement until the water level reaches the opposite deadband limit. The control mode for this type of Little-Man algorithm is called multistage, three position with antihunt.

The Little-Man control technique is applied to canal systems using the local automatic control method. The application of the Little-Man algorithm is most successful in implementing the constant upstream depth method of operation.

3-14. Colvin Algorithm

The *Colvin algorithm* is based upon the rate control mode in combination with the three position control mode. The Colvin controller has been used in many Bureau projects for the control of the outlet gates at diversion dams. It also has been applied successfully for local automatic downstream control for a single canal check structure having a short canal reach downstream. The Colvin controller can be implemented either with electromechanical or electronic components. The Colvin algorithm easily can be implemented using a microprocessor to execute the programmed algorithm. Referring to figure 3-13 (the electromechanical type of Colvin controller), usually the sensor consists of a float-tape-pulley-counterweight assembly. The comparator unit consists of a slip clutch, balanced bar, and two magnetically operated switches. The slip clutch is attached to the float pulley shaft. Then, the bar is attached to the slip clutch mechanism. A permanent magnet is attached at each end of the balanced bar. The magnets operate switches that are located at each end of the bar.

As the water level changes, friction of the slip clutch allows the bar to swing with the rotation of the float pulley shaft. When the bar swings through an arc distance—equivalent to the vertical distance of the

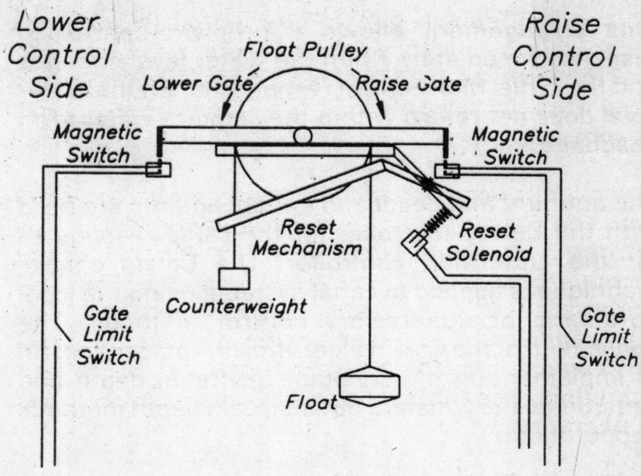

Figure 3-13. — Colvin controller.

water level deviation—the permanent magnet operates the switch. The switch contact closure energizes the control element. The control element consists of a run timer and the null position reset solenoid. The timer energizes the raise/lower relay or *actuator* to operate the motor and move the gate for a preselected time interval. At the same time, the spring-loaded reset solenoid is energized. The reset solenoid returns the balanced bar back to the null position by overcoming the friction of the slip clutch. The operation of the control element is referred to as *set-operate-time/variable-rest-time* (SOT/VRT).

By using rate of change in water level to determine the gate movements, rest time between gate movements becomes a variable. The rest time will be shorter if the water level changes rapidly, and longer, if the water level changes slowly. The VRT function permits the control action to be proportional to rate of change of water level. In most applications of the Colvin controller, a set of two gate run timers are added to the actuator element. The timers provide flexibility to select the proper gate run time to match the anticipated disturbances. Because the rest time is proportional to the rate of change of water level, the ability for the controller to achieve a match between the supply and demand of the canal system is excellent.

The control action of the Colvin controller can allow the water level to stabilize away from the desired setpoint water level. This reaction is characteristic of rate type control and reset action is often required to move the water level to the target level. The Bureau has used a Little-Man controller with the Colvin controller to provide the reset action. The Little-Man controller is modified to operate on a delayed basis so that the Colvin control rate action provides the primary response to the disturbance.

This arrangement allows the Colvin control to respond immediately when the water level changes and the Little-Man control responds when the water level does not return within the setpoint water level deadband.

The antihunt and deadband enhancements are used with the Colvin controller for the same reasons as for the Little-Man controller. The Colvin control technique is applied to canal systems using the local automatic or supervisory control methods. The application of the Colvin algorithm is most successful in implementing the constant upstream depth and sometimes the constant downstream depth methods of operation.

3-15. EL-FLO and EL-FLO + RESET

The EL-FLO (electronic filter level offset) and EL-FLO + RESET algorithms are based upon the proportional and proportional-plus-reset control modes. Proportional-plus-reset mode of control offers a great deal of versatility and operational flexibility to automatic flow regulation in canal systems. Large and rapid, as well as the small and slow, changes of canalside demands can be regulated smoothly by the proportional control mode. Operational stability is maintained by the *electronic filter* element. The reset control eliminates the residual water level offset of the proportional control and maintains a nearly constant downstream water level for all canal steady-state conditions.

The EL-FLO algorithm is applied to the automatic downstream control of canal systems. Downstream control concept requires that the water level be measured at the downstream end of the pool and the discharge into the reach is through the check structure gates at the upstream end of the pool. The control algorithm can be implemented either with analog type electronic components or with a microprocessor to execute the programmed algorithm. Referring to figure 3-14, one sensor is located in the downstream pool; it consists of a float-driven potentiometer that converts water level to an electrical voltage. The other sensor converts the check gate position to an electrical voltage using a potentiometer driven by the hoist mechanism. The water level voltage value is filtered by an electronic low pass filter and the output of the filter is

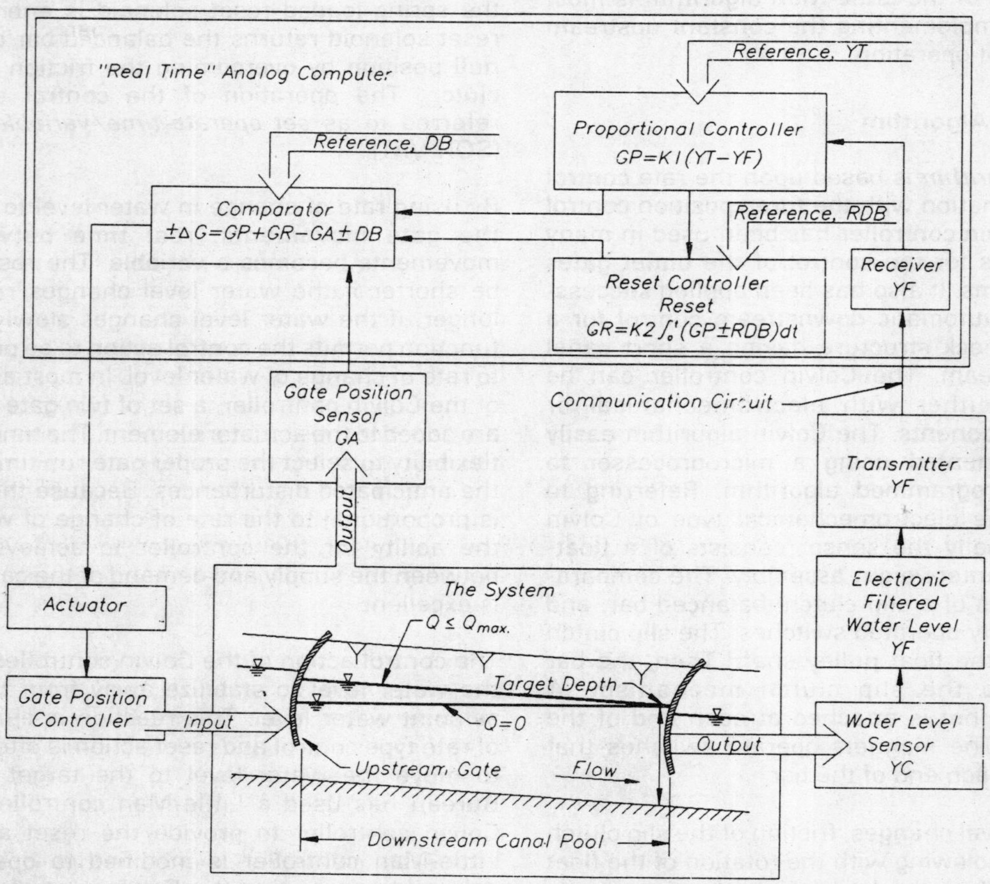

Figure 3-14. — EL-FLO controller.

transmitted to the upstream check structure where the control element is located.

The control element performs the proportional and reset calculation based upon filtered water level input, gate position input, and setpoint or target water level. The control element also consists of raise and lower relays that operate the motor actuator. Raise and lower relays are operated based upon results of the proportional and reset calculations performed by the control element. The actuator drives the motor to move the check gate. The desired gate position is determined by the algorithm and the gate position sensor input confirms that the gate moved to the calculated position.

The EL-FLO or EL-FLO + RESET techniques require a communication circuit between the downstream end of the pool and the upstream check gate location. The communication circuit is used to transmit the downstream filtered water level to the control element. These techniques also require monitoring of check gate position. Setpoint or target water level is the desired value to be maintained for all steady-state flow conditions. A deadband is used for gate operation to prevent short rapid gate movements.

Selecting the proper filter time constant and control parameters is essential to the correct and stable operation of the EL-FLO or EL-FLO + RESET feedback control system. These parameters should be selected based upon the results of mathematical model studies. Selection by using trial and error methods is difficult and usually results in poor control.

The EL-FLO and EL-FLO + RESET control techniques are applied to canal systems using the local automatic or supervisory control methods. The application of the EL-FLO or EL-FLO + RESET algorithms is most successful in implementing the constant downstream depth method of operation. Additional information on the EL-FLO can be found in reference [2].

3-16. P+PR

The P+PR (proportional plus proportional reset) algorithm is based upon the same control modes as the EL-FLO + RESET algorithm. The primary difference between the P+PR and EL-FLO + RESET techniques is that the P+PR algorithm is applied to the automatic upstream control of canal systems. The upstream control concept requires that the water level be measured just upstream of the check gate structure. The P+PR algorithm can be implemented with either the analog type electronic components or a microprocessor to execute the programmed algorithm. Referring to figure 3-15, one sensor is located just upstream of the gate; it consists of a float-driven potentiometer that converts water level to an electrical voltage. The other sensor converts the check gate position to an electrical voltage using a potentiometer driven by the gate hoist mechanism.

Upstream water level value is filtered by an *electronic filter* that eliminates wind waves and other influences of short disturbances from causing gate movements. The output of the electronic filter is used as an input to the *control* element. The control element performs the proportional and reset calculation based upon filtered water level, gate position, and setpoint or target water level. Control element consists of raise and lower relays that operate the motor actuator. The raise and lower relays are operated based upon proportional-plus-reset calculations performed by the control element. The actuator drives the motor to move the check gate. The desired gate position is determined by the algorithm and the gate position sensor input confirms that the gate moved to the calculated position.

Setpoint or target water level is the desired value to be maintained for all steady-state flow conditions. A deadband is used for gate operation to prevent short rapid gate movements.

Selecting the proper control parameters should be performed using mathematical modeling of the canal system as described for the EL-FLO control technique. The P+PR control technique is applied to canal systems using local automatic control or supervisory control methods. The application of the P+PR algorithm is most successful in implementing the constant upstream depth method of operation. Additional information on the P+PR controller can be found in reference [3].

3-17. BIVAL

The BIVAL algorithm, patented by the French company SOGREAH, requires simultaneous measurements of upstream and downstream water levels within the canal pool. BIVAL can be applied to the constant volume or the constant upstream depth methods of operation, but usually it involves a compromise between these two methods. BIVAL control has been most effective when the pool water surface pivots about a point slightly upstream of the pool midpoint. Coefficients in the simple linear control equation are adjusted to provide rapid volume change within each canal pool.

The BIVAL technique is applied to canal systems using the local automatic or supervisory automatic control methods. Additional information on the BIVAL algorithm can be found in reference [4].

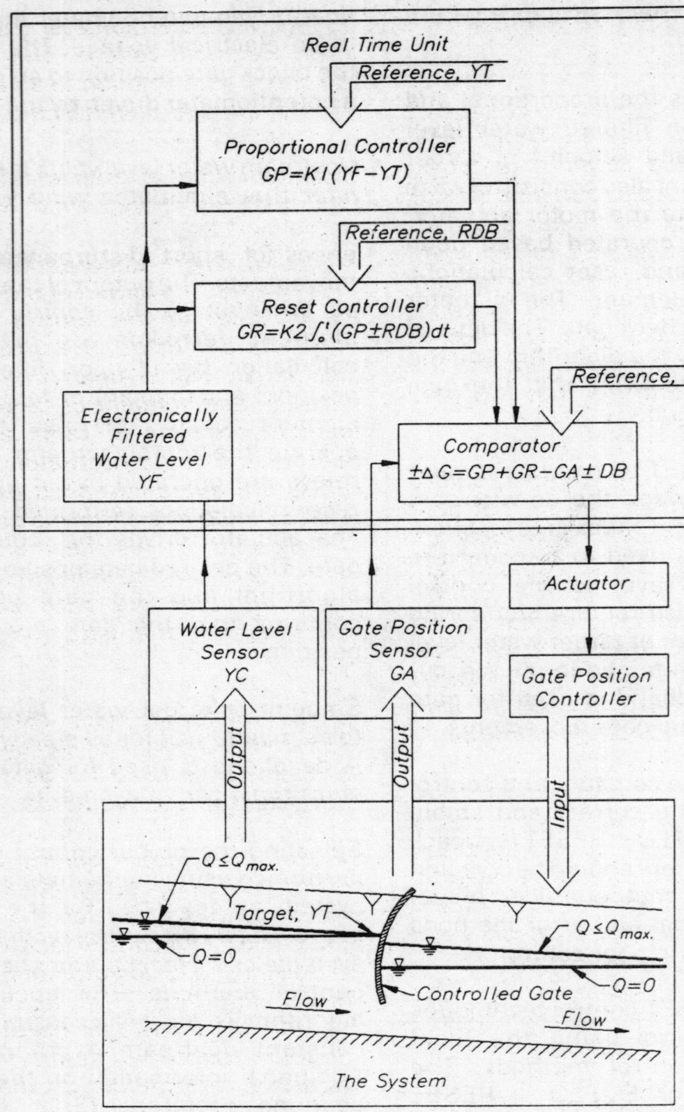

Figure 3-15. — P+PR controller.

3-18. Controlled Volume

The *controlled volume* (dynamic regulation) method of operation is implemented using many different algorithms. This technique requires that the canal system be controlled by the supervisory automatic or supervisory computer-directed control methods. The supervisory control method must be used because canal control is being performed based upon the real-time dynamic conditions of the canal system and the complexity of operation would require frequent intervention by operators. Usually, the controlled volume method of operation requires a mathematical model of the canal system to estimate the required gate positions, pumping plant flow requirements, and turnout requirements for the canal demand conditions during each hour of canal operation.

Control structures and pumping plants are controlled based upon conditions in the entire canal system—not just those conditions in the vicinity of controlled structure. Supervisory control provides the data acquisition system to collect the required data from the entire canal system. Then, these data are used in algorithms and in the mathematical model to dynamically adjust all the system control structures to meet the required canalside demands.

Canals and pumping plant irrigation management systems that use the controlled volume method are being operated successfully using the supervisory control method and special control algorithms. Examples are the Central Arizona Project in Phoenix, Arizona, and the Canal de Provence in France. Additional information on these projects and on the controlled volume method can be found in reference [5].

3-19. Gate Stroking

The *gate stroking* technique can be used to control transients in a canal subjected to large variations in discharge or water levels. Operation requires a mathematical model of the canal that is designed to continuously calculate required gate position based upon the scheduled delivery, depth, and discharge values specified at the end of the pools. The gate stroking algorithm calculates gate operation schedules to meet the depth and discharge values specified. The gate motions determined by this method can be quite irregular if the flow changes are large.

A supervisory control system is required to implement gate stroking. Because all gates must operate at a precise time, and by the correct amount, a centralized control system must be used. Gate positions can be transmitted from the master station to each check structure RTU and the gate operation schedule can be stored in the RTU for execution.

The gate stroking technique has been applied to canal operation. A description of this application and additional information about the method can be found in reference [6].

3-20. Variable Target

The *variable target* technique uses a changing setpoint or target to compensate for changes in demand or storage. The depth setpoint can be intentionally raised to provide storage of surplus water when demand decreases. Alternatively, the setpoint can be lowered to release surplus water when the demand increases.

Using supervisory control, the variable target technique is applied conjunctively with one of the other canal control algorithms. Setpoint changes are issued, from a master station, based upon system-wide conditions. After a setpoint change, normal control resumes. Whatever control algorithm is normally used will automatically adjust the water depth to the new depth setpoint.

This technique is useful when unused canal capacity exists which can be used for water storage. However, because of varying water levels, variable target is limited in application. For instance, it cannot be used where gravity turnouts require a constant canal water depth. Drawdown restrictions for lined canals also must be observed.

AUTOMATIC CONTROL COMPONENTS

3-21. Canalside Controllers and RTUs

The automation system may use local automatic control equipment or supervisory control equipment. The local automatic equipment provides local control of a control structure, with its communications limited to alarming and to remote reading of some sensor measurements. Local automatic equipment must be supervised periodically by ditchriders to determine the condition of the equipment and proper operation. Failures are reported using the alarm system and this information is used by the watermaster (at the headquarters) location to dispatch ditchriders to the troubled location.

Local automatic equipment can be assembled using mechanical parts to perform the automatic control according to the desired algorithm. Most commonly, local automatic equipment is assembled from electronic components and usually consists of microprocessor type equipment. The microprocessor equipment can execute the desired algorithms and perform necessary logic calculations for the automatic control. Typical canal control equipment is comprised of:

- Microcomputer
- Keyboard
- Digital display
- Analog and digital input and output equipment
- Communication channel interface
- Canal control algorithms (programs)
- Manual interface equipment
- Software operating system
- Developmental hardware and software systems

These controllers are easily maintained and programmed. An algorithm can readily be selected for compliance to the desired canal operation.

The software operating system is usually stored on the nondestructive memory of the microcomputer. This memory is called read only memory (ROM) and can be permanent, electrically erasable, or erasable under ultraviolet light. The contents of this type of memory will not be erased during power failure or when the microcomputer is "powered down." Real-time data such as water levels, gate positions algorithm constants, targets, and gains are stored in random access memory (RAM). Oftentimes, data in RAM is retained using a battery backup system to maintain parameters through short power outages. This memory is dynamic and is changed as required by the control algorithm. The automatic control equipment is installed in weatherproof steel or fiberglass cabinets to protect it from vandalism and the harsh environment associated with canal systems.

When. used with supervisory control, the local automatic controller must be microcomputer based and is called an RTU. The RTU consists of a microcomputer with the same capabilities as the local automatic controllers except the system will include interface equipment to a communication system. The RTU receives and interprets commands from the master station and executes the commands by operating the desired control structure device. The RTU also monitors the devices at the control structure location and sends this information to the master station on a periodic basis. If the RTU is performing automatic control of the canal control structure, the information that is being monitored and the control action that is being performed can be observed by the operator at the master station. Figures 3-16 and 3-17 show the configuration of a typical RTU or canalside controller.

3.22 Master Stations and Control Centers

The supervisory control system consists of a master station, RTUs, and a communication system. The RTUs monitor the canal control structures (check gates, pumping plants, turnouts) and send the monitored information to the master station over the communication system. A master station is shown on figure 3-18.

Master station equipment consists of a minicomputer or microprocessor based system including:

- Computer
- RAM memory
- Disk drive
- Keyboard
- Communication channel interface
- Hard copy device(s)
- Color video display unit
- Software (computer programs)

A color video display unit is used to present information to the operator at the master station. Video display units provide the operator with a vast amount of data on the entire canal system using a precise system of display screens. The display screens also are used to provide the operator with

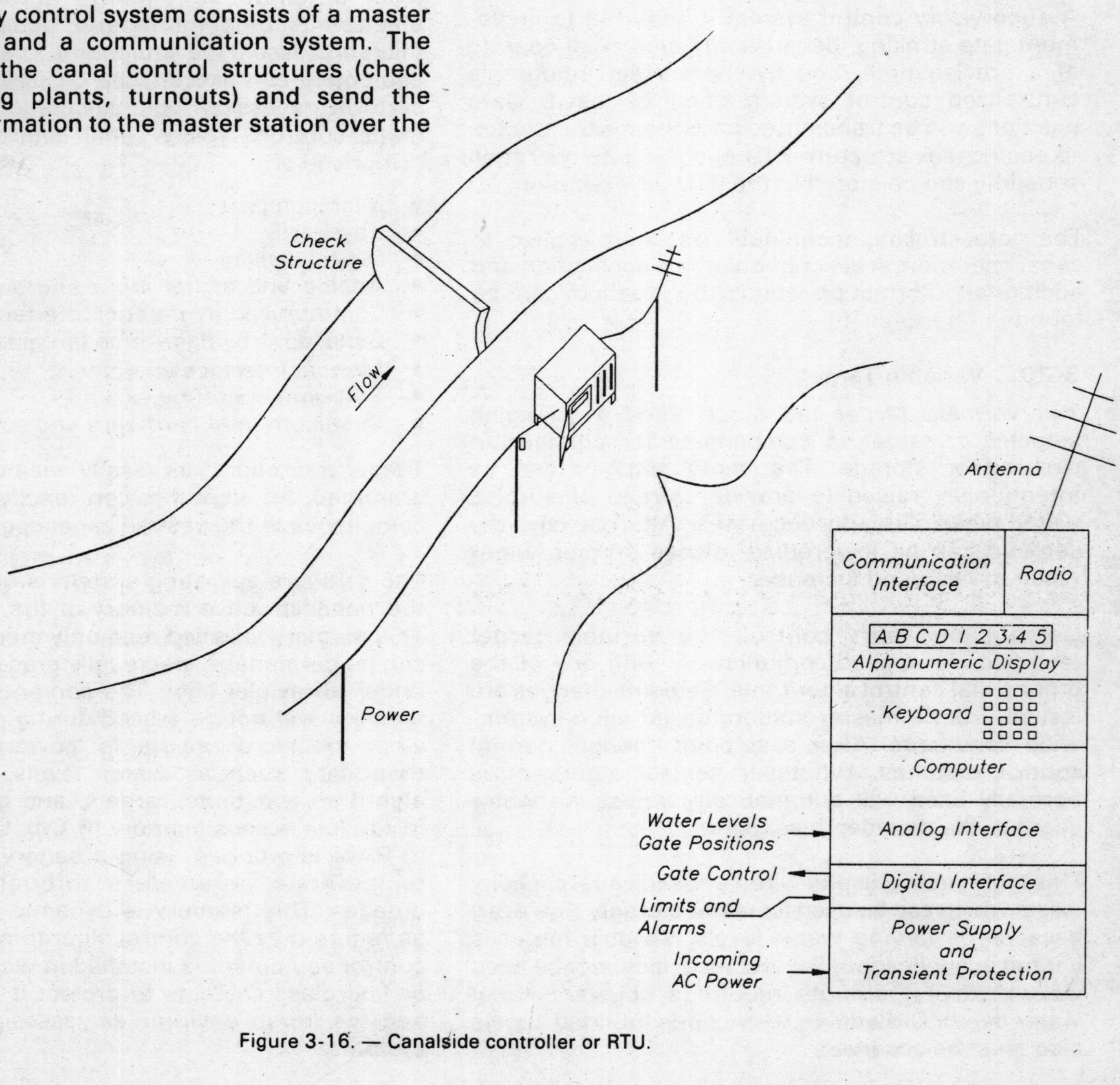

Figure 3-16. — Canalside controller or RTU.

Figure 3-17. — View of computer based RTU.

the proper procedures to control various devices in the canal system. The video display unit can present information to the operator using lists, or using graphic symbols and charts. The master station will include a hard copy device for printing event and alarm logs.

Supervisory control equipment provides for control of the canal system from a central location. This enables one operator to supervise the operation of an entire canal system from the master station. The operator can determine canal system conditions and make adjustments to the control structures through the RTUs using the communication network. The operator is informed of alarm conditions by monitoring data from the RTUs and the communication network.

When the operator is involved in all of the decisions concerning canal control, the canal system is in supervisory manual control. Supervisory manual control can be relatively simple and reasonable in cost. When RTU equipment is automatically controlling a canal control structure, with minimal operator intervention, the canal system is in supervisory automatic control. When the master station equipment automatically determines the operation of the RTUs the canal system is in supervisory computer-directed control. Supervisory computer-directed systems are complex, require special programming, and are expensive.

Additional information on master station and RTU equipment can be found in references [7 and 8].

3-23. Sensors

Automated canal systems use sensors at the control structures to measure water levels, flows, and gate positions. Many different types of sensors are

Figure 3-18. — Typical master station control console.

available for each type of measurement; it is important to select sensors that are best suited for operation on a canal system. Sensor selection requires consideration of several factors:

- Accuracy
- Reliability
- Resolution
- Repeatability
- Calibration requirements
- Temperature range
- Maintainability
- Availability
- Cost

Successful operation of automated canal systems depends heavily upon the performance of the sensors. Sensors can malfunction and cause the automatic control system to perform undesirable and catastrophic canal system operations. An absolute necessity is that automatic control equipment be designed to detect abnormal sensor readings to prevent the controller from taking improper control action.

Sensors can be classified as either analog or digital. This classification is based upon the output signal of the sensor. If the sensor output signal is a voltage, current, or frequency it is considered to be an analog sensor. If the sensor output is represented by binary digits or direct-current pulses it is considered to be a digital sensor.

Digital sensors are more accurate and less sensitive to electrically induced errors because the output is in a binary format and can be converted easily to engineering units at the automatic control equipment location. If a microprocessor controller is used, the conversion of the sensor output to engineering units is a simple digital process performed within the controller. However, digital sensors are more expensive, more complex, more difficult to repair, and often more prone to failure.

Analog sensors are less expensive, easier to maintain and repair, and easier to calibrate. Generally, analog sensors are more reliable over long periods of time. The output signal is an electrical analog signal that is susceptible to interference and degradation before it reaches the automatic controller. If the controller is microprocessor based, the analog signal must first be converted to a digital signal before the controller can use the data for control. This conversion is performed at the controller or RTU and may not be as accurate as a digital signal because of the transmission errors that may have occurred between the sensor and controller. The transmission errors can be reduced to a minimum when proper precautions are considered in the installation and wiring. An analog water level sensor is shown on figure 3-19. Additional information on sensors can be found in reference [9]

3-24. Communication Systems

Selecting a communication system for canal system automation requires careful consideration of these criteria:

- Canal system method of operation
- Location of the facilities to be automated
- Reliability requirements of the canal system operation
- Lifetime communication system cost
- Existing communication systems

Each criteria must be evaluated before the appropriate communication system can be selected. The design, installation, and quality of maintenance of the communication equipment will determine the overall success of the canal automation project.

Figure 3-19. — Analog water level sensor.

Several types of communication systems are available that are well suited for canal automation projects:

- Single channel VHF radio (30 to 300 MHz)
- Single channel UHF radio (300 to 3000 MHz)
- Microwave (3 to 30 GHz)
- Metallic cable (4 to 100 pair)
- Fiber optic cable (duplex)

Selecting the type of communication system that will best satisfy the needs of the canal automation project depends on the following:

- Control method to be used
- Location of master station
- Number of local automatic controllers or RTUs
- Data points at each controller or RTU location
- Data acquisition collection and transmission rates (number of points per second)
- Control output requirements of the controller or RTU
- Canal system reliability requirements
- Canal system configuration
- Cost

Each of the above items must be evaluated for a particular project before the best communication system can be determined. The control method must be selected before the type of communication system can be determined. Sometimes, an existing communication system is available and can be used for canal automation communications. The possibility exists to use an existing communication system for both voice and data transmission.

Based upon the Bureau's experience, the most versatile communication method for canal automation projects is the direct burial, multipair, metallic cable system as shown on figure 3-20. The security of

Figure 3-20. — Installation of buried cable.

transmission, flexibility of system configuration, installation cost, yearly maintenance, and ease of use account for this type of communication system being the most desirable for canal system automation projects. Additional information on communication systems for canal applications can be found in reference [9].

BIBLIOGRAPHY

3-25. Bibliography

[1] Johnson, C. D., "Microprocessor Based Process Control," Prentice-Hall.

[2] Buyalski, C. P., and E. A. Serfozo, "Electronic Filter Offset (EL-FLO) Plus Reset Equipment for Automatic Control of Canals," Bureau of Reclamation Report No. REC-ERC-79-3, 145 pp., Denver Colorado., June 1979.

[3] Buyalski, C. P., "Study of an Automatic Upstream Control System for Canals," Bureau of Reclamation Report No. GR-78-4, 79 pp., Denver, Colorado., August 1978.

[4] Chevereau, G., and M. F. Gauthier, "Use of Mathematical Models as an Approach to Flow Control Problems," Proceedings of the International Symposium on Unsteady Flow in Open Channels, BHRA-IARH, Newcastle-Upon-Tyne England, pp. J1,1-12, 1976.

[5] Gooch, R. S., and A. L. Graves, "Central Arizona Project Supervisory Control System," Proceedings of Computerized Decision Support Systems for Water Resource Managers: Case Studies of Large Systems, ASCE Spring Convention, Denver, Colorado, 1985.

[6] Falvey, H. T., and P. C. Luning, "Gate Stroking," Bureau of Reclamation Report No. REC-ERC-79-7, 86 pp., Denver, Colorado, July 1979.

[7] Solicitation/Specifications, 7-SP-60-02686/DS-7715, "Programmable Master Supervisory Control System, West Oakes Test Area, Pick-Sloan Missouri Basin Program, North Dakota," Missouri Basin Region, Bureau of Reclamation, Billings, Montana, 1987.

[8] Solicitation/Specifications, 7-SP-80-03880/DS-7750, "Programmable Master Supervisory Control System, Dolores Project, Colorado," Upper Colorado Region, Bureau of Reclamation, Salt Lake, Utah, 1988.

[9] Burt, C. M., "Overview of Canal Control Concepts," *Planning, Operation, Rehabilitation, and Automation of Irrigation Water Delivery Systems*, ASCE, edited by D. D. Zimbelman, pp. 81-109, New York, 1984.

CHAPTER 4

CONTROL SELECTION FOR EXISTING CANAL SYSTEMS

PLANNING

4-1. Planning Process

The planning process will lead to the proper application of practical and effective control systems. The process ends with the prototype control system installed, tested, and functioning on the actual canal system. Control selection should follow several important phases of the overall planning process.

The first phase in the planning process is *project definition*. Control selection planning begins with a description of project objectives and physical characteristics which define the project. Operating criteria are established to create an efficient match between the canal system's capabilities and its requirements. Physical constraints that will interfere with project objectives are identified. All of this pertinent prerequisite information should be carefully specified before proceeding.

The second planning phase is *control method selection*. This phase should begin with selection of operation and control concepts that will satisfy the operating criteria. (Control concepts, control methods, and methods of operation are described in chapters 1 and 2.) After the appropriate concepts have been identified, the control method and method of operation can be selected. Usually, more than one control method and method of operation will satisfy the operating criteria. Often, operation studies are necessary during this phase. Alternate control methods may have to be analyzed using computer models before selecting the best method.

The third phase of the planning process is the *definition of requirements*. An existing canal system may require structural modifications and additional equipment to accommodate a change in control method. This phase identifies preparation details such as site investigations and equipment considerations. Additional operation studies usually are required to define control strategies and operational parameters.

The fourth phase evaluates the *feasibility* of the selected control method. Feasibility is based upon the estimated costs of the new control system and the benefits achieved.

The final phase, *implementation*, establishes final design details for onsite preparations and control equipment. Specifications are written for procurement, installation, and acceptance testing of the control system equipment. Onsite modification may be required to interface with the new control system. Then, control equipment is installed, calibrated, and tested. To ensure the equipment will operate as designed, testing is required before and after installation. After it becomes operational, equipment must be properly maintained to minimize outages.

The following sections discuss each phase of the planning process in detail.

PROJECT DEFINITION

4-2. Project Description

Project description begins with identifying the objectives of the canal system. The basic canal system objective is to convey water from the source(s) to areas of use downstream. A delivery system is identified with irrigation, municipal and industrial, and fish and wildlife objectives. However, the nature of the water source, the type of conveyance, and the delivery concept will specifically identify the objectives of the canal project. These factors also will determine whether the canal system will be *demand-* or *supply-oriented*.

Canal systems are classified as either *delivery, collector*, or *connector* (see ch. 1). Each classification is identified for primary water use objectives:

- The *delivery* system is identified as one conveying water to downstream turnouts. Usually, a delivery system is *demand-oriented* and has many points of diversion along the system.
- The *collector* system is identified as one which increases project water supply and decreases flood damage. A collector system is *supply-oriented* and may have any number of sources.
- The *connector* system is identified as one which conveys water from one source to a single point downstream. Usually, connector systems are identified with intakes to pumping plants, powerplants, and regulating reservoirs. A connector system can be either *demand* or *supply* oriented. Typically, connector systems do not have multiple diversion or collection points.

A canal system's delivery concept determines the response requirements of the main canal. (Chapter 2 discusses delivery concepts.) The delivery concept is based upon the level of freedom the individual water users have as to the time and quantity of water received. Delivery freedom ranges from severe restrictions (rotation), to limited flexibility (scheduled), to unrestricted use (demand). The main canal must respond according to the delivery concept employed, from occasional and predictable flow changes to frequent and unannounced flow changes.

Most canal project objectives will become evident when the nature of the water supply, the type of conveyance, and the delivery concept are described. Some less apparent objectives may exist that also must be properly identified. For example:

- The canal system may be required to convey off-season surplus water into storage reservoirs.
- The system may include pump/generating plants.
- Pumping may need to be maximized during offpeak energy hours and minimized during on-peak hours to reduce energy costs.
- Variances in turnout diversion rate may be dependent on critical water scheduling and irrigation requirements.

Describing the project's characteristics and physical properties helps identify the canal system's capabilities to satisfy its objectives. A schematic profile provides a brief but concise project description. An example is shown on figure 4-1.

The schematic profile provides a visual image of the entire canal project. The following basic prerequisite information can be readily identified:

- The characteristics of the water supply source.
  - Is it gravity or pumped?
  - Is the canal headworks from a reservoir, diversion dam, another canal, or from a natural river channel?
  - Is the upstream reservoir storage system extensive and does it include generating/pumping plants?
- The locations of the canal check gate structures.
  - How long are the canal pools?
  - How many check gate structures are involved?
  - How many gates are at each check gate structure?
- The maximum design discharge for each canal pool.
- The location of canalside turnouts.
  - In which areas will demand be high and low?
  - What is the type of turnouts: gravity or pumped?
  - Are the turnouts to be automated?
  - What is the maximum design discharge for each turnout?
- The location and flow capacity of wasteways, inverted siphons, drop structures, or other structures which affect flow.

The above are examples of the types of information that the schematic profile can show to easily identify important operational characteristics and physical properties. Additional project information must be studied to identify possible operational problem areas.

For example, a canal headworks may divert from an afterbay (tailwater) channel below a powerplant. Powerplant afterbay water levels can vary significantly as flow releases fluctuate to meet power needs. Frequent adjustments will be required at the canal headworks to maintain a constant water supply to the downstream canal system. Therefore, the canal headworks must be monitored and adjusted frequently.

Another example is a branching canal system like that on figure 4-1. Branching canals are difficult to operate manually. The problem is one of matching the supply to the demand of each branch. Minor unannounced variations in canalside turnout demands can cause significant water level variations before operators can make the necessary upstream

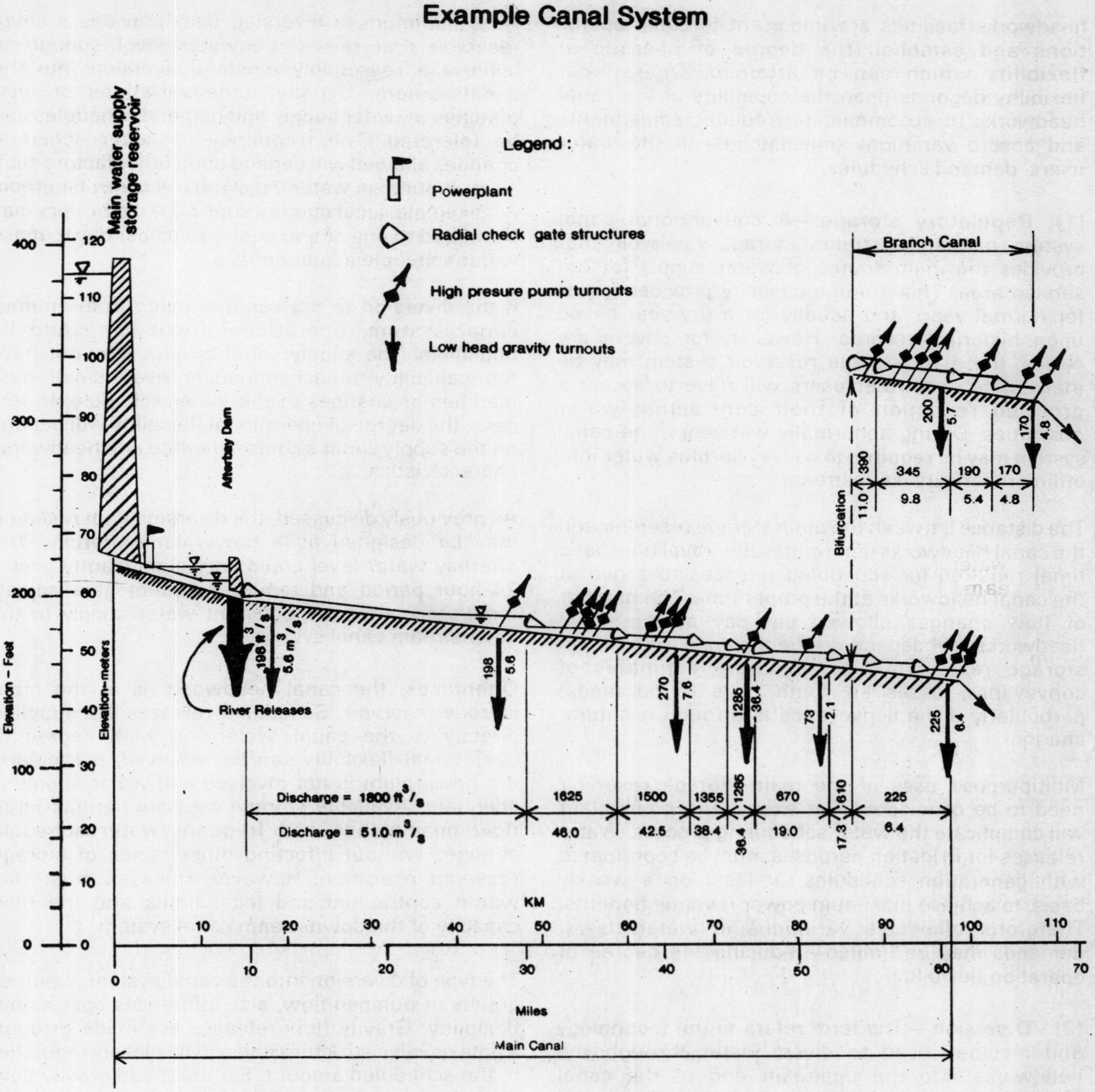

Figure 4-1. — Schematic profile of a canal system to identify objectives and capabilities of a project.

supply flow adjustments. Frequent monitoring and adjustment of the two branches is required. Even then, difficulty is encountered maintaining water levels within reasonable limits.

The preceding are just two examples of the types of operating criteria that need to be identified and evaluated based upon the physical description of the canal project facilities. The following discusses, in more detail, basic requirements of a project's physical description.

The canal system consists of three major parts as follows.

a. **Canal headworks.**—The headworks are the facilities used to divert water into the upstream end of the canal system. The characteristics of the headworks facilities are important to canal operations and establish the degree of operational flexibility which can be attained. Operational flexibility depends upon the capability of the canal headworks to accommodate frequent adjustments and absorb variations (mismatches) in the water users' demand schedules.

(1) Regulatory storage.—A conventional canal system has an upstream storage reservoir that provides the main source of water supply for the service area. This storage capacity provides water for normal years and usually for a dry year based upon historical records. However, for severe dry cycles, the main storage reservoir system may be inadequate and water users will have to accept a prorated reduction of their contractual water quantities. During abnormally wet years, the canal system may be required to convey surplus water into offline reservoirs downstream.

The distance between the main storage reservoir and the canal headworks determines the travel time (lead time) required for scheduled releases to arrive at the canal headworks at the proper time. The number of flow changes allowed per day at the canal headworks will depend on the distance to the main storage reservoir. Also, accurate estimates of conveyance losses en route have to be made, particularly if the conveyance is through a natural channel.

Multipurpose uses of the main storage reservoir need to be considered. For example, a powerplant will complicate the water scheduling process. Water releases for irrigation purposes must be coordinated with generation schedules, at least on a weekly basis, to achieve maximum power revenue benefits. Therefore, allowable variations in water users' demands may be limited—reducing the degree of operation flexibility.

(2) Diversion.—The term refers to the technology and facilities used to divert water through the headworks into the upstream end of the canal system. If the canal is supplied from a natural channel, a diversion dam usually is constructed to create a small reservoir to supply the canal. The regulatory storage capacity and the allowed range of water level fluctuations in the reservoir are critical to the operational flexibility of the canal system.

The ideal diversion dam design would have enough regulatory storage to provide the highest degree of operational flexibility. However, the diversion dam site location usually is not ideal—water supply is inconsistent—and the cost of an ideal design is prohibitive. Therefore, the diversion will likely impose some restrictions on canal operation.

As a minimum, a diversion dam provides a small reservoir that raises the water level enough to achieve a reasonably efficient diversion into the canal system. Usually some variations or mismatches in water supply and demand schedules can be tolerated. The frequency of water schedule changes allowed will depend upon other factors such as multipurpose water uses and the control method. For example, local manual control by ditchriders may be limited to one or two onsite visits per day to make water schedule adjustments.

If the diversion to the canal is made from another canal system, operational flexibility could be negligible. The supply canal system may not have the capability to accommodate even small mismatches or changes in the water schedule. In this case, the degree of operational flexibility will depend on the supply canal's control method and headworks characteristics.

As previously discussed, the diversion dam reservoir may be designed as a powerplant afterbay. The afterbay water level could vary significantly over a 24-hour period and require frequent gate adjustments to maintain a constant water supply to the downstream canal system.

Oftentimes, the canal headworks is at the main storage reservoir. Scheduled releases are supplied directly to the canal system. A high degree of operational flexibility can be achieved, particularly if a powerplant is not involved and water supply is adequate. Available storage capacity permits large flow mismatches and frequent water schedule changes without affecting other facets of storage reservoir operation. However, releases must stay within contractual and legal limits and the flow capacity of the downstream canal system.

The type of diversion into the canal system, whether gravity or pumped flow, also influences operational flexibility. Gravity flow releases are made through a gate or valve structure and can be closely regulated to the scheduled amount. For low-head gravity flow diversion, regulation is greatly affected by small variations in the upstream and downstream water levels. High-head gravity flow diversion is less sensitive. Where the diversion comes directly from the main storage reservoir, a high-head pressure gate or valve with a downstream energy dissipator may be required. Small variations in the upstream and downstream water levels will not greatly affect the diversion quantity.

Both low- and high-head type gates and valves require accurate calibration of gate opening versus discharge, which is dependent on upstream and downstream water levels. Gate calibration determines the capability to maintain accurately scheduled diversions. Errors will cause flow mismatches and affect water levels in the canal system downstream.

A headworks pumping plant typically delivers flow in increments equal to the smallest pump unit capacity. However, the pumping plant could include a bypass facility where surplus flow—the total pump unit flow less the total canalside turnout demand—is returned to the pumping plant intake. Pump units having variable speed motors provide a better match between pumped supply flow and total canalside turnout demand.

Operational flexibility is reduced if pumping power requirements have to be scheduled in advance with the power company. Penalties may be assessed if pumping differs from the announced schedule, particularly if pumping is increased. Additionally, pumped diversion is always subject to power outages or individual pump unit failures that will cause sudden and unannounced decrease in flow to the downstream canal system.

b. Main conveyance channel.—Main conveyance facilities include the canal prism, check gate structures, and other flow control structures. Physical properties of these structures establish constraints that limit the flexibility of operation.

Open channel flow is difficult to regulate. Flow changes take the form of waves which travel at relatively slow speeds. The wave velocity and time of travel are dependent on the canal prism geometry and the canal pool length. Therefore, the description of physical properties begins with a typical cross section of the canal prism shown as on figure 4-2.

Cross-section properties include geometry and hydraulic properties for all canal reaches. This includes a description of freeboard allowance; it is indispensable for predicting the canal pool storage regulation potential. The available freeboard is a primary factor in selecting the canal pool method of operation. The physical properties of the canal prism, particularly the canal pool length, may limit the selection of control method that can be applied to the canal system.

Canal pool length determines the degree of operational flexibility of the main canal system. The longer canal pools will be subject to larger water level fluctuations as the canal flow changes from

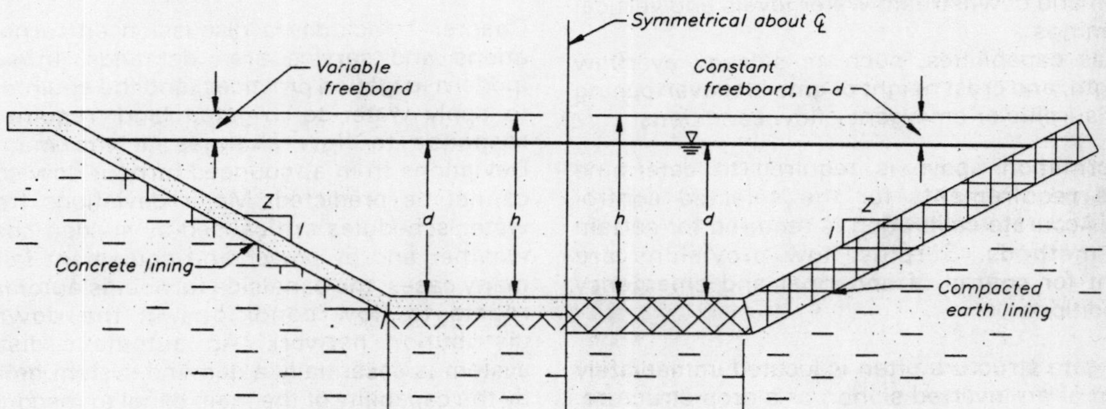

Canal section, Station	Area A	Velocity V	Discharge Q	Hydraulic radius r	Friction coef. n	Bottom slope S	Side slope Z	Bottom width b	Maximum depth d	Canal lining height h or var. El.
Sec. 1, 0–41+00										
Sec. 2, 41+00–97+10										
Sec. 3, 97+10–515+20										
Sec. 4, 515+20–640+07										

Figure 4-2. — Example of a typical canal prism physical and hydraulic properties.

one steady state to the next. The larger water level fluctuations may affect constant delivery requirements at low-head gravity flow turnouts. As discussed (see ch. 2) the rate of flow change may have to be limited to avoid exceeding drawdown criteria for the longer canal pools. Larger flow changes can be accommodated in the shorter canal pools without exceeding drawdown limitations. Thus, operational flexibility increases for the shorter canal pools; i.e., the travel time decreases and the system responds more quickly to a new steady state.

The canal check gate structure is the next important main conveyance channel canal feature. The check gate structure maintains upstream water levels based upon the method of operation and regulates the flow into the downstream canal pool according to the water schedule. Physical properties that need to be described are:

- Locations of check structure
- Number of gates and type—radial or vertical lift
- Types of gate hoist power—electrical, hydraulic, or manual
- Gate hoist motors—constant or variable speed
- Calibration of flow versus gate opening as a function of upstream and downstream water levels
- Staff gauges for visual measurements of upstream and downstream water levels and vertical gate openings
- Bypass capabilities, such as sidebay overflow weir length, and crest height or gate leaf overtopping flow, for surplus or emergency flow conditions

The information above is required to determine interface requirements for the selected control method. Accurate calibration is required for certain control methods. Surplus flow provisions are important for control of abnormal and emergency flow conditions.

A check gate structure often is located immediately upstream of an inverted siphon or a drop structure. Usually, inverted siphons are used to convey the main canal flow under natural cross drainage channels. Drop structures are used when it is necessary to lower the main canal invert elevation to accommodate sudden changes in topography. Siphons and drop structures can cause free flow conditions at the canal check gate located immediately upstream. Normally, the check gate is designed to operate under submerged flow conditions, wherein the flow is a function of both the upstream and downstream depth. When free flow conditions occur, the downstream water depth will not affect the flow calibration, so a different flow versus gate opening calibration as a function of only the upstream depth occurs. Therefore, operation of the downstream canal pool will not affect the flow at the upstream check gate under free flow condition.

Inline relift pumping plants will have the same physical characteristics and properties as described for the headworks pumping plant. Matching canal scheduled flow to the inline pumped flow will be difficult if the canal system does not have canal pool regulation storage or reservoirs. Inadequate inline storage is similar to a canal headworks pumping plant having a small diversion dam reservoir intake.

Other main canal features will need to be described if they impact on the operating criteria or change the main canal operation. Features affecting water level fluctuations or the regulation of mismatches should be considered. For example, an offline storage reservoir may be included to store surplus water during the flood season for release back during the irrigation season. Usually, pumping or pump/generating plants are involved with offline storage facilities. The location, capacity, and crest elevation of wasteways need to be identified.

c. **Canalside turnouts.**—Water is diverted through canalside turnouts from the main canal to the individual water users or to distribution networks. Turnouts are discussed separately because turnout flow demands and physical properties have a major influence on main canal operation.

Chapter 1 includes a discussion on turnout operations and service area demands. In summary, modern irrigation practices, and the equipment used to apply water to irrigated land, require a quick response to flow changes in the main canal. Deviations from announced turnout flow schedules cannot be predicted. Major deviations from daily water schedules are caused by sudden changes of weather and by power and equipment failures. In many cases, the canalside turnout is automated and reacts to flow conditions in the downstream distribution network. An automatic distribution system is essentially a demand system limited only by the capability of the main canal to respond. These operational characteristics have a major impact on the selection of the control method.

Additional operating characteristics influence the design of the distribution system downstream of the canalside turnout. A canalside turnout to a pipe distribution system will operate differently than one to an open channel. For a pipe system, whether supplied by gravity or by pumps, the canalside turnout gate is opened at the beginning of the irrigation season and remains open until the season ends. Therefore, the demand at the turnout is changed immediately when individual water systems turn on or off. For an open channel distribution system, operation of the canalside turnout must be coordinated closely with the main canal operation. Low-head gravity flow turnouts may be sensitive to changes in main canal water levels. This is a major factor that reduces operational flexibility, unless special provisions are incorporated into the turnout design.

Balancing the main canal operation is difficult for a canal with unattended canalside turnouts where the gates remain open. Actual turnout diversions are unknown because the turnouts are not controlled by main canal operators. Canalside turnouts that regulate water diversion may require occasional or frequent operator attendance depending on the delivery concept. Therefore, the frequency of flow changes and type of gate control at canalside turnouts are important operating criteria if the balance between main canal supply and demand is to be precise all the time.

4-3. Operating Criteria

Operating criteria for a canal system should include the objectives, the constraints that limit operational flexibility, and the improvements required to upgrade the operation.

a. Objectives.—The canal project description includes detailed objectives of the headworks, the conveyance channel, and the canalside turnouts. The objectives serve as a basis for selecting the operating and control method. Therefore, operating criteria should begin with a brief but concise summary of the objectives.

b. Constraints.—Constraints limit the canal system capability to respond to changes in canalside turnout demands and maintain an efficient supply/demand relationship. Constraints reduce operational flexibility and affect the ability to maintain the desired water levels and flow schedules in the canal system. If water levels and flow changes are physically limited, flexibility to adjust for normal, abnormal, and emergency conditions is constrained.

All the constraints discussed in chapter 2 can limit operational flexibility. Some of the major constraints that need to be included in the operating criteria are summarized as follows:

(1) Water level fluctuation.—The range of water level variation, between the minimum and maximum levels, will limit the operational flexibility of a canal system. Water levels in the canal prism must be maintained above the minimum level for canalside turnouts to deliver the maximum design flow rate. Conversely, water levels should not exceed the maximum designed depth of the canal prism for normal steady-state flow conditions. These limits on water level variation establish the potential canal pool regulatory storage for normal operations. Canal lining provided above the maximum depth is reserved for freeboard which protects the canal embankment from wind wave erosion. However, higher water levels during the unsteady-state period following a normal flow change, or caused by abnormal or emergency flow conditions, are temporarily permitted to encroach into the freeboard. Water level should never be allowed to overtop the canal lining—even under emergency flow conditions. Severe erosion can develop and cause the embankment to fail in a short time period. Therefore, lining height determines the degree and duration of abnormal or emergency conditions that can be safely accommodated.

The rate of decrease in the water level is limited by drawdown criteria, as discussed in chapter 2. This limitation affects the acceptable rate of flow change in the canal pool.

Locations in the canal system where water level variation will affect flow regulation need to be identified. Water level fluctuation at low-head gravity flow control structures, such as canalside turnouts, can cause adverse operations downstream. A small variation in the water level could cause a significant change in the flow rate—upsetting the balanced flow conditions downstream. If water level needs to be maintained constant all the time, flexibility of operation will be significantly reduced at these locations.

Canal check gate structures often are designed with sidebay overflow weirs (sometimes referred to as wing walls, overflow walls, or bypass weirs). A sidebay overflow conveys surplus water into the downstream canal pool when high water level occurs. Weir freeboard is typically small, which may prevent sudden flow reductions at the check structure. Sidebay weirs may also prevent using some methods of operation, such as constant volume or constant upstream depth, that require greater than normal depth upstream of check structures at low flows.

(2) Canal pool flow regulation.—The principal flow control structure of the canal system is the *canal check gate structure*. Check gate flow adjustments are based upon the canal water schedule. Any factors that may affect flow adjustment and consequently limit operational flexibility need to be identified. For example, existing gate hoist motors may not have been designed for frequent gate position adjustments.

The check gate spacing may result in long canal pools. The canal pool length can impose two important operational flexibility constraints (1) recovery time, and (2) water level drawdown criteria.

The combination of pool length and canal prism depth determines the time delay required to adjust the canal pool water surface profile from one steady-state flow to the next. Therefore, longer canal pools require more time to recover to a new steady-state. Also, the change in water surface elevation per unit of flow change will be greater in the longer canal pools. Accordingly, flow changes that decrease the water surface elevation rapidly may have to be limited to avoid exceeding drawdown criteria.

The number, location, capacity, and type of *canalside turnouts* need to be identified. Gravity flow turnouts require adjustments to adhere to the water schedule, and the diversion rate will change if the main canal water surface elevation varies. The diversion rate at pumped turnouts may vary in large flow change increments equal to the pump unit capacity, and as a result, cause a cyclic unbalanced flow condition in the canal pool. Usually, turnout diversion rates will not match the announced schedule as accurately when individual water users operate their own turnouts [1]. This upsets the balanced operation of the canal pool. Automatic pressure pipe distribution systems are troublesome for canal operations because they can create frequent unscheduled flow changes.

Relift pumping plants usually inhibit pool flow regulation. Bypass capability or variable speed pump motors allow the pumping plant discharge to match the downstream water schedule, but if these features are not provided, one pump unit may be required to cycle *On* and *Off*. If automatic start and stop capability is provided, details such as the *On* and *Off* water levels and pump unit capacities need to be known. Also, provisions for pump unit lockout, sequencing, and start/stop delay timing need to be determined.

(3) Scheduled power.—Power must be scheduled for the larger pump/generating plants to optimize revenue and power consumption. Therefore, water users may be required to submit weekly water demand schedules in addition to their daily schedules. Limited schedule modification by the water users reduces the flexibility of operation. The objective to optimize power revenue and cost may require on-peak operations for power plants and offpeak operation for pumping plants whenever possible. Delivery flexibility may be sacrificed to meet these power objectives. The larger relift pumping plants will require power to be scheduled one day in advance of the actual pumping. Therefore, additional pumping to meet unscheduled increases in canalside turnout demands cannot be accomplished without incurring a penalty cost.

(4) Regulatory storage.—Available regulatory storage capacity of the canal system establishes operational flexibility to balance significant variations between supply and demand schedules. Off-peak pumping is practical when regulatory storage on the canal system is large. Storing off-season surplus water in storage reservoirs is also possible. In a canal system having little or no storage capacity, water users must adhere closely to their announced schedules.

(5) Wasteways.—Wasteways provide a safe method to divert surplus flow from the canal when storage capacity is not available downstream and the canal inflow exceeds the storage capacity. The location and flow capacity of wasteways should be included in the operating criteria. If wasteways cannot be used or are not provided, then the canal must respond immediately to abnormal or emergency flow conditions.

(6) Inverted siphon.—As stated in section 4-2b, an inverted siphon may cause free flow conditions to occur at the canal check gate located at the siphon inlet. A different gate opening is required when free flow occurs compared to the submerged condition at the same flow rate. It is possible for flow to oscillate from free to submerged conditions at the same gate opening, which results in variable flow rates to the downstream canal pool. Frequent gate adjustments would be required to maintain a constant flow rate. The flow and water level conditions that cause the free to submerged oscillating flow conditions need to be identified.

c. Improvements.—Areas where improvements can be made to upgrade the canal operation become evident when the operating criteria is properly summarized. Flow control structures that require frequent adjustments are costly to operate manually. A balanced operation and good service to the canalside turnout is difficult to maintain. However, water users will benefit if the operation of the canal could be more flexible, particularly when their crops are young and sensitive to climatic changes. The low-head gravity turnouts sensitive to main canal water level fluctuations should perhaps be modified or provided with a local automatic controller so that constant diversion (as scheduled) can be maintained.

The delivery characteristics may indicate that sudden and unannounced canalside turnout changes are the norm rather than the exception. Immediate response on a continuous basis is required to prevent surplus or deficient flow from propagating into the lower pools of the canal system.

A canal system without provisions to divert and store surplus flows is subject to costly damages if prompt controlling actions are not initiated. Pumping plants are always subject to complete shutdown—rejecting flow into the canal prism. These and other abnormal and emergency conditions require prompt and deliberate attention.

Relift pumping plants always reduce the flexibility of canal operation, particularly with large pump capacities. However, with proper scheduling and efficient use of available regulatory storage on the canal system, operational flexibility can be maximized. To reduce energy costs, it may be possible to implement offpeak pumping operations.

CONTROL METHOD SELECTION

4-4. Operation and Control Concepts

After all operational criteria have been established, the proper operation and control concepts can be applied as described in chapter 1 and elsewhere. The concepts must be compatible with the established operating criteria.

a. **Operation concept.**—Operating criteria specify if the canal system is demand-oriented or supply-oriented. The examples just discussed are demand-oriented for normal flow conditions. Individual water users schedule their downstream demands (within contractual amounts) which determines the flow schedule required at the canal headworks, at the flow control structures of the main canal, and at the canalside turnouts en route. Canals classified as delivery systems usually are demand-oriented (downstream operation concept). However, delivery systems can experience periods of water supply shortage or surplus when the operation must be based upon the upstream concept.

Individual water users must accept a prorated reduction in their contractual amounts during a water shortage. The reduced amounts match the available upstream water supply at the canal headworks. For some canal systems, available water is habitually in short supply even though the canal is classified as a delivery system. In this case, the operational concept is normally upstream. The operation reverts to the downstream concept during wet years or during periods when the water supply is sufficient to meet the water users' downstream demands.

Canal operation becomes supply-oriented when surplus water is conveyed to offline reservoirs. The water schedule for the flow control structures is based upon the upstream water supply available at the canal headworks during a water supply surplus period.

The collector type of canal system is always supply-oriented. Available water supply inflows at the headworks and into the canal pools en route determine the water schedule for the downstream flow control structures.

Therefore, selecting the correct operational concept (downstream or upstream) is a matter of applying operating criteria to determine if the canal system water schedule is demand- or supply-oriented. Also, operating criteria indicate if the water schedule orientation will change based upon the nature of the water supply conditions.

b. **Control concept.**—Selecting either the downstream or upstream control concept is simply a matter of being compatible with the downstream or upstream operation concept. The applicable control concept is an important factor in applying the correct control method. The basic information used to adjust the flow control structure according to the water schedule must come from either the upstream or the downstream direction and is not interchangeable (figs. 1-4a and 1-4b).

4-5. Methods of Operation

Several methods are available to change flow to satisfy the water schedule requirements of the canal system. As discussed in chapter 2, the common methods of canal pool operation are:

1. Constant downstream depth
2. Constant upstream depth
3. Constant volume
4. Controlled volume

Changing the flow involves adjusting the canal pool water surface profile and check gate openings. All three check gate operating techniques—sequential, simultaneous, and selected—can be used to achieve each of the methods of canal pool operation.

Each method of canal pool operation and each check gate operating technique has its own advantages and disadvantages, as described in chapter 2. Control method selection involves evaluating the characteristics of the different methods of operation. The intent is to apply methods of operation that are compatible with the operating criteria specified above. Later, each method of operation will be discussed further as it applies to each available control method.

The choice of the canal pool method of operation and check gate operating technique depends upon the following considerations:

- Freeboard allowance and canal pool length constraints
- Response and recovery characteristics

a. Freeboard and pool length constraints.—The primary constraint that determines the appropriate method of canal pool operation is the provision for canal freeboard. Canal embankment and lining freeboard is a major capital cost. The minimum freeboard required, to protect against embankment erosion from water surface wind waves, would be constant and parallel to the water surface profile for the maximum design flow as shown on figure 4-3a.

For constant and minimum freeboard, the only practical method of canal pool operation is the constant downstream depth method of operation. The depth, Y_{max}, is held nearly constant at the downstream end of the canal pool for all steady-state flow conditions, as shown on figure 4-3a. Water fluctuations are minimal at this location, which is an advantage for major canalside turnouts located at the downstream end of the canal pool.

However, the storage wedge on figure 4-3a requires a time delay in check gate operations to adjust from one steady-state flow condition to the next. Changes in canalside delivery flows also may be delayed. Long canal pools exhibit long time delays when establishing a new steady-state. Therefore, canal pool length is another physical constraint that determines the response time required to adjust canal flow. The canal headworks may need to have regulatory storage or a lead time to adjust the canal pool storage wedge as the steady-state flow changes.

When freeboard is constant and minimum, the other methods of operation available for canal pool operation can only be used at low flows. As the canal pool water schedule approaches the maximum design capacity, the method of canal pool operation must approach the constant downstream depth method of operation.

The constant upstream depth method of canal pool operation requires the maximum freeboard, as shown on figure 4-3b. The cost of the additional embankment and lining is the main disadvantage of this method of operation. However, the time delay required to adjust the steady-state flow usually is not a constraint on operations because it does not delay the canalside delivery for demand-oriented canals. Headworks storage is usually not required because additional storage is provided within the canal pools.

The variable freeboard shown on figure 4-3c enables the constant volume method of canal pool operation. Additional lining is required for the downstream half of the canal pool length to allow the water surface profile to pivot near midpool for all steady-state flow conditions. Because the canal pool volume remains

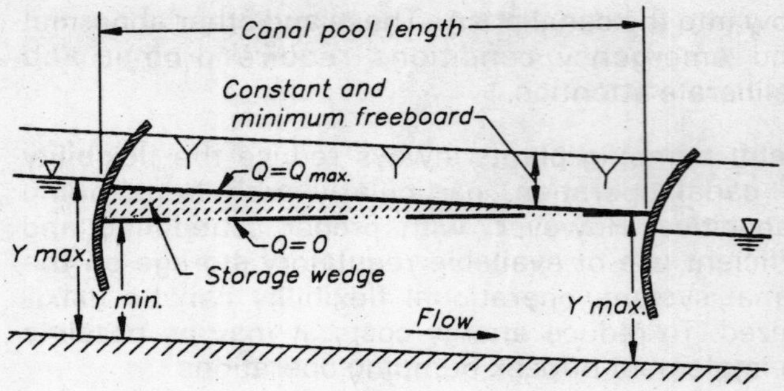

(a) Constant and minimum freeboard designed for constant downstream depth method of canal pool operation.

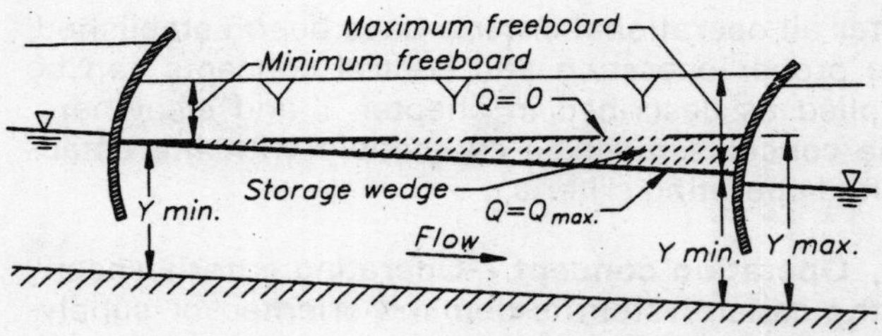

(b) Maximum freeboard designed for constant upstream depth method of canal pool operation.

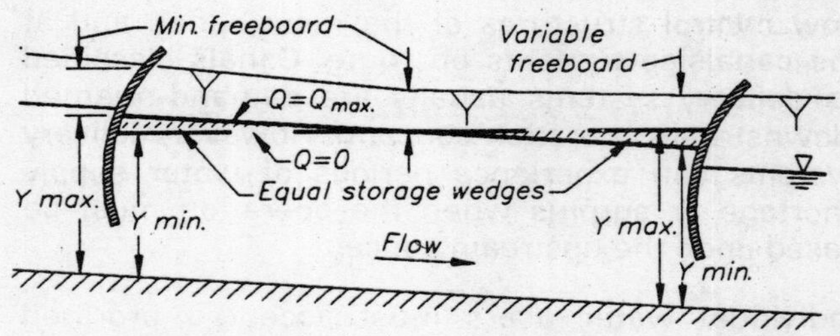

(c) Variable freeboard design for the constant volume method of canal pool operation.

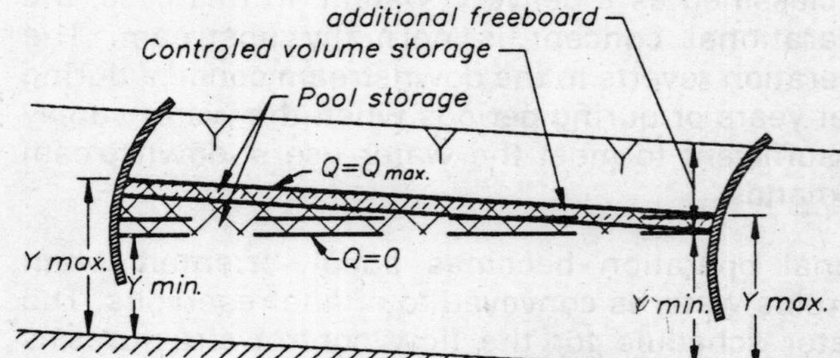

(d) Constant and additional freeboard designed for canal pool reregulatory storage and controlled volume method of canal pool operation.

Figure 4-3. — Canal freeboard allowance.

constant for all steady-state flow conditions, a time delay between check gate adjustments is not required. Also, regulatory storage at the canal headworks is unnecessary.

The controlled volume method of canal pool operation requires additional canal lining freeboard throughout the entire length of the canal pool as shown on figure 4-3d. The additional lining is used for canal pool reregulation storage to better match the water supply and demand schedules. The amount of additional lining freeboard is based upon the degree of mismatch.

b. Response and recovery characteristics.—Immediate response to normal water delivery schedule changes and to abnormal or emergency flow conditions, in addition to rapid recovery to a new steady-state flow condition, are the ideal characteristics of canal pool and check gate operation. However, ideal characteristics are difficult to achieve. The response characteristics are dependent on the selected check gate operating technique. The recovery characteristics are dependent on the method of canal pool operation.

Sequential check gate operating technique has a slow response characteristic because of the time delay involved. Usually, the time delay is equal to the time required to adjust the storage wedge to the new steady-state water surface profile, which is dependent on the length of the canal pool. *Simultaneous check gate operation* is the best technique to respond immediately to normal, abnormal, or emergency flow conditions. However, the recovery to steady-state still requires a time delay to adjust the canal pool water surface profile. *Selected check gate operating technique* is best for correcting minor mismatches when they occur.

Constant downstream and upstream depth methods of canal pool operation have slow recovery characteristics because of the time delay required to proceed from one steady state to another. Normally, the constant depth methods of operation are accomplished by the sequential gate operating technique. The water level fluctuations from one steady state to the next are the greatest at the end of the canal pool opposite from the constant depth end. In long canal pools, particularly those with relatively large canal invert slopes, flow changes will have to be limited to avoid exceeding drawdown criteria at the opposite end of the canal pool.

Constant volume operation has the best response and recovery characteristics of all the methods available for operating canal pools. The simultaneous check gate operating technique responds rapidly while maintaining a constant pool volume. Recovery within a single pool will take only about one-half the time required for the constant downstream and upstream depth methods of operation. More importantly, constant volume operation can respond and recover in all canal pools at the same time, regardless of the length of the canal system. A canal may take all day to recover from flow change with a constant depth method of operation, but less than an hour with the constant volume method of operation.

The controlled volume method of operation will have variable response and recovery characteristics depending on the check gate operating technique used and on the amount of volume change.

The physical description, operating criteria, and the correct application of concepts and methods of operation have been noted. The descriptions have established the capabilities, limitations, and operating requirements of the canal system. The appropriate control method can now be selected.

4-6. Selecting the Practical Control Method

When selecting a control method, the goal is to optimize the operation within the constraints placed on the canal system. Upgrading the operation—providing a better match between the canal system conveyance capabilities and the canalside demand requirements—will increase water conveyance and water use efficiencies.

The control method must be compatible with the control concept of the canal system. The two basic control concepts, downstream and upstream, are described in chapter 1. The control concept is dictated by the location of the information needed for control relative to the control structure.

Adjustments to the canal system's flow control structures can be accomplished by four basic control methods:

1. Local manual
2. Local automatic
3. Supervisory
4. Combined

Control methods were introduced in chapter 1 and described in chapter 3.

The application of each control method has advantages and disadvantages. Costs and benefits vary significantly from local manual to supervisory control. The control system that offers the most advantages may be the most expensive to implement. At this stage, elimination of inappropriate or impractical control methods is emphasized.

Typically, more than one control method will satisfy the control concept and specified operating criteria; appropriately, alternative control methods should be considered. The final selection should be based upon economic feasibility—the cost versus the associated benefits—considering operational flexibility and overall project objectives.

The following describes applications of the four basic control methods and their associated advantages and disadvantages.

a. Local manual control.—Local manual control involves a team to operate the canal system flow control structures onsite. Team effort and practical experience is required to achieve a successful operation. Each team member must use judgement in their respective areas of responsibility.

The water schedule (developed by the watermaster and other team members) is based upon the downstream or upstream operational concept; i.e., the water schedule priority is either demand- or supply-oriented. The ditchrider does not need to know the operational concept. The ditchrider's responsibility is to change the canal flow to satisfy the water schedule.

Traditionally, ditchriders have implemented the constant downstream depth method of canal pool operation to satisfy the water schedule. Check gates are adjusted by the sequential technique progressing in the downstream direction. The selected check gate operating technique is used on the ditchrider's return trip, progressing in the upstream direction, to make minor adjustments if necessary.

(1) Applications.—Local manual control is the most basic canal control method. This control method can be used with all methods of canal pool operation and all check gate operating techniques. However, simultaneous check gate operation is impractical because it would require an operator to be stationed at each check gate structure when flow changes are made.

(2) Advantages.—Local manual control is probably the most economical control method for simple canal systems with a single purpose water use. The operating team can develop skills through experience. Usually, highly technical expertise is not required.

Personal contact and cooperation between the ditchrider and the individual water user provides intangible benefits. Water users, who actually pay for the canal project, have a part in the overall canal system operation. Personalized service can accommodate minor changes in canalside diversion when critical water applications are needed for young crops.

Frequent ditchrider coverage of the canal physical features can detect problems such as vandalism, right-of-way encroachment, and potential embankment failure. Based upon experience, a ditchrider can use judgement to anticipate and make decisions in the field when abnormal or emergency conditions develop.

(3) Disadvantages.—Local manual control has many disadvantages. Time is required to gain experience and to develop the necessary skills for efficiently operating the canal system. For example, the calibration of gate flow versus gate opening is developed over a period of years using data visually observed for different flow conditions. During the learning period, canal operation may be inefficient.

Local manual control restricts water schedule changes at canalside turnouts to one or perhaps two changes per day. Water users must adhere to their announced schedules even when other factors such as farm activities and weather changes may warrant a change in operation. The implementation of modern irrigation practices may be limited when canalside diversions must be constant for long periods between visits by the ditchrider.

Abnormal and emergency flow conditions may not be detected immediately by the ditchriders. When an abnormal or emergency condition is discovered, necessary adjustments upstream and downstream of the problem area take longer. Severely unbalanced flow conditions frequently cause either high or low water levels in the canal prism because of the inherent slow response characteristics of local manual control. Slow response is a particular disadvantage when modern irrigation practices are used in the distribution networks by individual water users. The slow response of local manual control is compounded for long canal systems.

The prime disadvantage of local manual control is the inability to optimize the operation of complex canal systems having multipurpose water uses or pump/generating plants. Many ditchriders are needed to operate complex canal systems. Efficient coordination of pump and generating schedules with water supply and demand schedules requires immediate response. During abnormal and emergency flow conditions, the local manual control method may not be responsive enough to prevent the occurrence of damaging high or low water levels.

b. Local automatic control.—Local automatic control consists of an arrangement of mechanical, electrical, and electronic components located at the control structure. The equipment automatically executes control algorithms to sense canal conditions and to make the necessary adjustments at control structures. Onsite visits and duties of the ditchrider are minimized [2, 3, 4].

Local automatic control equipment can range from simple, inexpensive electromechanical devices to complex and more costly computer installations. The simpler electromechanical devices are limited to simple single purpose tasks, whereas, electronic/computer local automatic control systems are capable of performing complex local control functions.

(1) **Applications**.—Whenever a control structure needs frequent adjustments, local automatic control is an ideal alternative to the local manual control method. Local automatic control can be designed to maintain water levels or flow rates within specified limits without the need for frequent ditchrider visits.

The automatic sump pump is an example of local automatic control. A simple electromechanical device automatically starts and stops the pump motor at specified high and low water limits. Thus, damage from high water levels caused by ground water seepage can be prevented without the need for ditchrider attendance.

The capability of the On-Off sump pump automatic operation can be expanded by adding a three-position mode controller. The additional control action provides ideal application for unattended diversion dams and has been used successfully by the Bureau. This type of controller has many applications in both the downstream or upstream control concept for a single control structure such as the headworks at a diversion dam, a sluice gate, or a canalside turnout.

The three-position control action also has application to a series of check gates in a supply-oriented canal system which is operated by the constant downstream depth method of operation. At each check structure, the sensor is located immediately upstream of the check gate.

The On-Off or three-position control device cannot be applied to a series of canal check gates for the downstream control of an entire demand-oriented canal system. The length of the canal pool introduces a time delay needed to adjust the water surface profile to a new steady-state flow condition. The time delay requires a time lag compensator in the closed loop feedback path to maintain control stability. These complexities to the control algorithm require an electronic controller.

Local automatic control can be applied to either demand- or supply-oriented canal systems. It is simply a matter of locating the sensor downstream or upstream of the control structure, respectively. The distance between the sensor and the control structure determines the complexity of the required control technique and the equipment requirements.

Local automatic control can be designed to operate the canal pools by the constant downstream depth, the constant upstream depth, or the constant volume methods of operation. Applying local automatic control to the controlled volume method of operation is not practical unless the local automatic control is combined with the supervisory control method.

(2) **Advantages**.—Local automatic control can be designed to regulate the single flow control structure and the more complex canal system having a series of check gates. Properly designed control algorithms can respond quickly to frequent changes—small or large—in the canalside turnout demands. Unscheduled changes are automatically coupled to the canal headworks water supply. Ditchrider visits can be virtually eliminated from the normal operation. Service on demand can be achieved that is compatible with modern irrigation practices, providing benefits to the individual water users [2].

(3) **Disadvantages**.—The inherent response delay at the canal headworks to changes in turnout demands requires flexibility in the water supply. If the canal headworks consists of a diversion dam, the delay would not be a problem. However, if the reregulating storage capacity is small, large mismatches in the water supply/demand relationship may not be accommodated. Water users may have to limit their delivery flow variations.

Long canal pool lengths may limit the application of the local automatic downstream control concept. Long pools require larger water level fluctuations to respond and recover to the new steady-state flow condition. Large water level fluctuations can exceed drawdown criteria and disrupt canalside delivery during the transient state [3, 5]. Provisions to limit water level drawdown within specified criteria cannot be included in the local automatic controller.

Typically, local automatic control requires electronic control equipment to execute specially designed control algorithms. Mathematical model simulation studies, using various real-time flow situations, are required to develop and verify control parameters. Personnel with mathematical model expertise and canal operations experience are required. An electronic/computer technician is needed to install, calibrate, and maintain the automatic control equipment. The cost of providing the required expertise may be prohibitive for smaller canal projects [2].

The local automatic control concept cannot be changed from downstream to upstream without changing the sensor location. A technician would have to relocate the sensor (or electronically switch the location if both sensors are installed) if the operation concept changes from demand to supply. The technician also must change the control parameters stored in the control equipment, because different parameters are required for the different control concepts.

Local automatic control needs a dedicated one-way communication channel between the water level sensors and check structures for control of check gates. Providing this communication capability will be relatively expensive if the sensors are located far from the checks. Additionally, a one-way communication channel should be provided from each check structure to a central location for alarm signals. This is an important but costly item.

c. Supervisory control.—Supervisory control allows the watermaster to monitor the canal system and to adjust control structures from the central headquarters. The operating team's skill and field experience are accumulated at the central headquarters to achieve a more successful canal system operation.

Operational data are collected automatically at sensors throughout the canal system and transmitted to the central headquarters. Data are presented in a suitable format to be analyzed by the watermaster. Based upon this information and the watermaster's judgment, the necessary control actions are initiated from central headquarters.

(1) Applications.—Supervisory control is applicable on all canal systems and with each method of canal pool operation and all check gate operating techniques. The main limitation is the high level of technical requirements.

Supervisory control is impracticable for simple, single purpose canal systems that can be efficiently controlled by local manual or local automatic conrol methods. Supervisory control is required for complex canal systems that need prediction and optimization capabilities to achieve an efficient operation. With a supervisory control system, a canal can maintain a better match between the water supply and demand to provide greater operational flexibility.

The level of supervisory control can range from monitoring data to be used for manually implemented control actions, with frequent watermaster participation, to totally computer-directed control, with infrequent watermaster participation. Supervisory manual control is applicable when the operation is simple and the results of control actions are easily predicted, such as control of unattended diversion dams.

Supervisory automatic control with a dedicated computer is ideal for normal operations of the canal system that has complex delivery requirements. For normal flow conditions, the watermaster's participation is limited to initializing and making occasional minor adjustments to the automatic controller. However, the watermaster must use judgment and manually issue control commands when abnormal or emergency conditions occur.

Computer-directed supervisory control is not commonly used on canal systems. Special and customized computer programs are required that must include judgement capabilities. To predict accurately the future operations and to program judgment for all possible flow conditions is extremely difficult. Also, a large staff is required to develop, implement, and maintain the specialized computer programs. Therefore, the application of computer-directed control is usually limited to large, complex canals.

(2) Advantages.—Supervisory control has many significant advantages. One person can have concurrent information on the entire canal system. Hence, changes in one section of the system that will affect other sections can be noted promptly and appropriate adjustments can be made. When abnormal or emergency flow conditions occur, the ability to monitor the entire canal system can help to prevent spilling water or damaging project facilities.

Excellent response characteristics can be achieved using supervisory control because many control structures can be adjusted simultaneously. Instantaneous and safe shutdown of the canal system can be accomplished. Also, supervisory control can accommodate a change in the control concept without changing control equipment. The water supply/demand schedules can be matched for a wide range of canal operations, including offpeak pumping, using the controlled volume method of operation. Therefore, supervisory control increases the flexibility, reliability, and efficiency of operation [1, 6, 7].

(3) Disadvantages.—The primary disadvantages of the supervisory control method are initial cost and the need for higher technology. Control and communication equipment on the canal and in the master station can be expensive. Supervisory control requires a dedicated two-way communication channel between the canal and central headquarters. Sometimes, redundant communication channels are required to ensure reliability.

Typically, the operating entity does not have staff expertise to implement a supervisory control system. Highly trained personnel are needed for installation, operation, and maintenance. For computer-directed supervisory control, control software is particularly difficult and expensive to develop.

d. Combined control methods.—Certain combinations of the local manual, local automatic, and supervisory control methods are practical. Most canal systems have operating characteristics that are compatible with all three control methods. Therefore, combinations are appropriate and should be given serious consideration. Applying two or three control methods to the canal system is limited only by the inventiveness and the ability of those involved in the planning process to recognize practical combinations.

(1) Applications.—Most canal systems need daily patrol by a ditchrider to observe site conditions for safety and preventive maintenance purposes. Commonly, the ditchriders perform local manual control during their regular patrol duties. Even when the canal system uses local automatic control at the check gates, other control structures or turnouts may need adjustment. These structures could be scheduled for manual adjustment during regular patrol visits instead of being equipped with automatic controllers. This type of operation would require less control equipment on the canal system, reducing equipment and maintenance costs.

Combining local manual and local automatic control is ideal for a canalside turnout or a diversion dam outlet gate. For example, local automatic control could regulate a gate to maintain a constant diversion flow (target flow value) as the water levels upstream and downstream change. Whenever a flow change is scheduled, the ditchrider could visit the site and change the target value to the new schedule.

A system using local automatic and supervisory control could be applied where diversion flow changes frequently. The target value could be reset from the central headquarters by the watermaster instead of requiring frequent ditchrider visits to the site.

Another ideal combination is local automatic control of the canal check gates with supervisory manual control backup. Local automatic control could maintain normal flow conditions and respond to most abnormal conditions. Backup supervisory control is efficient when local automatic control equipment fails. Canal check gates upstream and downstream of a failed check gate controller could be adjusted manually from the central headquarters to maintain a balanced operation until the equipment is repaired. An emergency condition, such as flood flow entering the canal, can be corrected by the watermaster using the backup supervisory manual control method.

A common plan is to combine local manual control with a supervisory control system that only consists of remote data acquisition capabilities. For example, a remote diversion dam may require different release schedules when flash floods occur. By monitoring water levels and rain gauges or stream flow gauges upstream, the ditchrider can be dispatched to the site to make flow adjustments only when flow conditions warrant attention.

Remote monitoring provided without control capabilities, at the central headquarters, can be achieved at a reasonable cost within the operation and maintenance budget of smaller canal systems. By merely monitoring critical parameters, the watermaster can significantly increase the efficiency of a conventional operation.

(2) Advantages.—The combined control method employs the best characteristics of the other three control methods as they apply to the operating characteristics of the canal system. Usually, practical combinations of control methods will achieve more benefits to the project than an individual control method can provide.

(3) Disadvantages.—In some cases, the disadvantages of the combined control methods are compounded. For example, the complexity of local automatic control is increased when combined with supervisory control. However, other disadvantages will be negated. The difficulty of changing the local automatic control concept from downstream to upstream can be easily overcome by adding supervisory control backup.

DEFINITION OF REQUIREMENTS

4-7. Operation Studies

Computer simulation studies are required to optimize the operation of complex canal systems. Studies are necessary to evaluate and verify the different control methods and methods of operation for various flow conditions. A canal system without wasteways or reregulating reservoirs has little margin for error when flow is near maximum capacity. Control during abnormal and emergency conditions is critical. To minimize pumping costs, other techniques (e.g., gate stroking [8]), may be appropriate. The evaluation, selection, and development of complex control techniques requires analysis with mathematical models.

Operation studies should include:

- Selection of control parameters for local automatic controllers
- Development of supervisory control strategies and procedures for offpeak pumping, abnormal, and emergency operations
- Evaluation of the canal system design—such as the type and location of check gate structures, wasteways, and storage facilities—to improve response and recovery characteristics
- Evaluation of alternative delivery concepts, e.g., testing the effects of different turnout flow changes on the overall canal operations
- Development of mathematical models that predict water supply and demand and optimize pump/generating schedules for complex canal systems

A combination of technical expertise and practical experience is required to conduct operation studies properly and to develop the requirements and specifications for the control equipment.

4-8. Equipment Considerations

Control equipment is needed to adjust the flow to the prescribed water schedule within the limits of the control method selected for the canal system. Each control method has specific equipment requirements.

Local manual control requires basic equipment such as staff gauges and flow measurement devices. The ditchrider visually observes staff gauges to obtain the upstream and downstream water levels at the check gate structure. Based upon a calibration of flow versus gate opening, as a function of the upstream and downstream water levels, the ditchrider can determine the gate opening required for the new steady-state flow. The gate opening is changed by using the staff gauge to obtain the correct new gate position. Frequently, check gates are equipped with motorized gate hoist mechanisms.

Flow measurement devices are installed at appropriate locations in the canal and at all canalside turnouts. The ditchrider uses flow measurements to confirm the desired flow schedules. Also, flow measurements are used to bill the water user or the water district for the actual quantity of water delivered. The ditchrider needs a mobile radio to communicate with the watermaster at the central headquarters.

Additional equipment is needed when local automatic, supervisory, or a combination of control methods are implemented. The additional requirements usually include stilling wells, equipment shelters, depth and gate position sensors, and mechanical and electrical equipment. Communication channels are required between the sensors and check structures, and between the check structures and central headquarters. More complex control methods require computers to perform mathematical model simulation studies and to execute real-time monitoring and control software.

Definition of requirements for the local automatic and supervisory control methods includes site investigations, operation studies, control equipment, testing, and maintenance.

4-9. Site Investigations

Site investigations are required to evaluate the existing onsite equipment and determine additional equipment needs. Typically, the local automatic and supervisory control methods require the following onsite equipment:

Stilling wells of proper design and location are necessary to measure water levels accurately and to shelter the water level sensor equipment. Stilling wells should have valving to permit easy cleaning of the feeder pipes without disturbing the sensor. The stilling well should not be located too close to the canal check gate, or other flow control structures.

Weatherproof enclosures attached to the gate(s) hoist mechanism protects the gate position sensor equipment. A direct drive type gate position sensor requires a flexible shaft connection to the gate hoist to accommodate misalignment.

Check structures should have adjustable gates equipped with motorized gate hoist mechanisms. Therefore, electrical power service must be provided. The gate motor must be designed to withstand high ambient temperatures, frequent starts, and long duty cycles. Variable speed motors may be required for the more complex control methods. A typical gate movement speed is 1.0 to 1.5 feet per minute (0.3 to 0.5 m/min). The rate of movement for variable speed gates typically varies from zero to 1.5 feet per minute.

Mode switches must be installed at each control structure to allow onsite equipment to be operated in the *local*, *automatic*, or *supervisory* mode. Maintenance personnel can set the mode switch to local for protection against inadvertent operation when performing equipment maintenance.

Shelters protect all onsite control equipment from the environment and vandalism. A separate shelter is normally included for electronic equipment, and at times with air conditioning or heating to maintain a desired temperature. The supervisory control method requires space for control and communication equipment at the central headquarters.

A backup battery system for the monitoring and control equipment permits continuous operation during power failures. Complex canal systems may include backup engine generator sets with automatic failover (when power fails) to operate gate hoist motors during extended power outages.

Buried electrical cables and/or conduits connect the sensors and the gate motor actuator to the control equipment. Communication cables should be separated from alternating current power and control cables.

Radio communication systems include antenna masts or towers. Direct wire communications require lightning-protected terminal blocks within the control house; fiber optic communications require fiber termination panels.

4-10. Control Equipment

The basic requirements for control equipment are:

- Water level and gate position sensors typically require a resolution of 0.06 inch (1.5 mm).
- The more complex control methods use computer-based control equipment. Hardware and software must be properly designed and tested for reliable operation. Software must execute control algorithms and start/stop logic exactly to design specifications.
- Mechanical/electrical control equipment for the simpler control methods must be designed to operate continuously for long periods of time without failures.
- The more complex control methods should include the following fail-safe procedures:
  - When automatic control equipment fails, all control output must be inhibited
  - When a gate or pump unit fails to start or stop as commanded, the control outputs should be de-energized after a preset time
  - Protection against simultaneous operation of raise and lower motor actuators
  - Delays between alternate raise and lower outputs on single phase gate motor equipment
  - Isolation equipment to protect control equipment from lightning and power line transients
  - Power regulation transformers for noisy and unregulated primary power circuits
  - Lightning protection for all direct buried cable connecting the various onsite components
- Alarms should be activated by the following conditions:
  - High or low water levels
  - Gate not at setpoint position
  - Loss of primary power
  - Loss of communications
  - Vandalism and intrusion

4-11. Installation and Testing

The installation of the control equipment requires careful consideration. The location of stilling wells, sensors, instrumentation, and canal control equipment should be planned to allow easy access for operation and maintenance

All local, city, and State regulatory codes should be followed when installing electrical wiring and conduit systems. Direct buried wiring and cable installations should be designed so that access to the cable is not impeded for future testing and repair. All direct buried cable on or near cultivated farmland should be buried to a mainimum depth of 4 feet (1.22 m).

The local automatic control or RTU equipment should be located so that the control structure can be operated conveniently by the ditchrider or technician. Most sensor displays and manual controls are located on the local automatic control or RTU cabinet; the cabinet installation should be such that the canal and control structure are visible from the cabinet position.

The *local, automatic, supervisory* mode switch should be located near the gate control structure. The mode switch should be housed in a vandal-proof cabinet with a locking door to prevent inadvertent operation of the switch. The switch should have the three-mode positions clearly identified and the switch should be mounted to permit easy operation by the ditchrider or technician.

Control equipment must be tested *before* it is installed on the canal system to ensure that it performs as required. After installation, a performance test should be run to assure calibration, interfacing, and communications are correct and will perform as required.

4-12. Maintenance

Eventually, control system equipment will require maintenance, recalibration, and repair. A preventative maintenance program must be initiated upon installation. A good preventative maintenance program will keep equipment failures and outage times to a minimum. An inventory of spare parts and components will aid timely repair. Periodic testing and examination of equipment should be scheduled.

Qualified technical expertise and training are important basic requirements. Local automatic and supervisory control methods require technical expertise to implement and maintain a successful control system. Providing qualified personnel and necessary training programs cannot be over emphasized.

FEASIBILITY

The feasibility of a new control system is based upon an evaluation of costs and benefits. Also, costs should not exceed what the operating entity can reasonably afford, which is based upon the water users' ability to pay.

For each proposed control system alternatives, all costs attendant to control system implementation should be estimated. Benefits are more difficult to evaluate because many benefits may be intangible. Judgment and operating experience are required to estimate the benefits associated with a control system.

Often the tangible benefits—benefits that have a clearly defined dollar value—are less than the cost of the new control system. However, intangible benefits (those benefits whose dollar value is difficult to assess) will make control system implementation worth the endeavor. Usually, the feasibility of implementing a new control system depends in large measure on judgment.

4-13. Costs

The cost of two or more alternative control schemes should be evaluated before making a selection. Costs will be associated with each of the following stages of control system implementation:

- Operation studies
- Onsite preparation
- Control equipment
- Installation and testing
- Maintenance

Operation studies can be costly, depending on the amount of mathematical model simulation required. Some control schemes require extensive computer modeling to develop and to verify the selected control method and the control parameters for various flow conditions. Several mathematical models are available; some will execute on a personal computer. It may be costly to acquire an appropriate mathematical model and to learn to use it effectively.

Oftentimes, operational studies are based upon trial and error solution, thereby requiring a long time to study all of the various flow conditions and operating scenarios. If a canal project does not have the capability to conduct mathematical model studies, these services will have to be contracted. Therefore, engineering costs for selecting and verifying the proposed control scheme can be significant.

Onsite preparation costs include stilling wells, sensors, equipment shelters, updating mechanical/electrical facilities, communication channels, electrical protection equipment, electrical conduits, wire and cable, and other items needed to prepare the site for the control system. The existing canal system probably will require onsite modifications to satisfy the equipment requirements which were discussed previously. The current cost of available control equipment should be used to estimate the cost of the proposed control alternatives. Actual final cost will not be known until equipment is purchased.

Control equipment costs are estimated based upon the experience of the control system designers and current information available from equipment suppliers. Dedicated communication channels likely will be the highest cost component. The costs for the control equipment should include equipment training, documentation, and spare parts.

Installation and testing costs include installation, calibration, and field acceptance testing of all control system components. These work items must be completed before the control system can be used for actual operation. Both hardware and software installation should be considered. Training personnel assigned to operate the canal with the new control system should be included. Installation costs will depend on how much site preparation is required to interface the control equipment with existing equipment. Costs will increase if installation is not properly planned or if equipment is not tested before shipment to the site.

Maintenance costs include labor and replacement parts. Qualified technical personnel with electronic/computer expertise must be employed or retained so that equipment outages are kept to a minimum. Training may be required to familiarize maintenance technicians with software and hardware for the control and communications systems. It is essential to have a complete spare parts inventory to facilitate timely repair service.

For estimating purposes, the Bureau of Reclamation generally considers the maximum cost of a control system to be approximately 2 percent of the total capital cost of the canal project (excluding right-of-way costs). Sometimes, higher costs can be justified when the control system upgrade results in tangible benefits or specific savings which will offset the cost of the new control system.

Usually, one has difficulty justifying the cost of complex control systems for small and existing canal systems. Even when the control equipment cost is acceptable, other costs associated with the implementation process may preclude project feasibility. Site preparations, engineering cost to perform operational studies, and maintenance costs associated with the requirement of greater technical expertise can easily negate any apparent benefits for smaller canal systems [2].

4-14. Benefits

The objectives of the canal system prescribe the control system benefits such as (1) better service to water users, (2) efficient water conveyance, and (3) reduced operating costs. Many benefits are intangible. Estimating monetary value for providing better service to the water users is difficult. An upgraded control system allows the water user to receive the scheduled quantity of water at the specified time, as opposed to the limitations of conventional control methods. When irrigators are able to apply the proper amount of water at the proper time, farm production and efficiency increase. Therefore, water users can receive significant benefits.

Efficient water conveyance from the source to the service area is essential to maintain water conservation objectives; routinely wasting water is no longer acceptable. Many water users compete for surplus water rights. Based upon the cost of supplying water, however, monetary savings of reduced waste will often be minor. On many projects, reducing waste water would add little to the economic justification of the control system. The benefits for these projects are primarily intangible, but are significant when rationalized that additional land could be irrigated by conserving water. Whenever water has been pumped, reducing waste will result in economic savings from the reduced power consumption. This can be a major tangible benefit on a project involving a significant amount of pumping.

Most existing canals, particularly the older systems, have wasteways. A control system has the potential of eliminating the need for wasteways. Intangible benefits result from the nonuse of wasteway channels, eliminating damage to wasteway channels and other adjacent properties.

The reduction of operation and maintenance costs is the main tangible benefit that can be compared to the cost of the control system. The number of ditchriders needed to control the canal system (particularly a complex system) can be reduced when local automatic, supervisory, or a combination of control methods are implemented. However, a minimum daily patrol of the canal system by ditchriders is necessary to check for encroachment, vandalism, or other potential hazards that could jeopardize safety. Therefore, smaller canal systems usually will not realize a significant personnel reduction. As a rule, the savings of eliminating two ditchriders will approximately equal the cost of adding one qualified technician to maintain a new control system.

Substantial cost reductions can be realized when pump/generating schedules are optimized. Implementing offpeak pumping operations, when possible, reduces energy costs. Optimizing generation increases revenue. The reduced cost for pumping energy and the increased power revenue can offset the cost of a control system.

A control system has potential to eliminate extreme high and low water level fluctuations caused by abnormal and emergency flow conditions. Therefore, excessive pressure changes and high water levels that cause damage to the canal lining and embankment can be reduced. The lower maintenance cost would be a tangible benefit and could be estimated based upon historical maintenance records.

4-15. Evaluation

The decision to implement a control system should not be justified solely on tangible benefits. Intangible benefits, as previously described, are significant on most canal projects. Estimating tangible benefits and properly describing intangible benefits related to the need to upgrade the canal operation is essential when estimating the feasibility of a proposed automation system.

The following guidelines help to evaluate the control system:

- Simple control methods are for simple tasks and the associated tangible benefits are relatively high.
- Complex local automatic control methods are for canal systems whose requirements are difficult to manage; they have the highest intangible benefits.
- Supervisory control provides system-wide control of project facilities; it optimizes operation for complex canal projects. Supervisory control has the most tangible and intangible benefits.

Management philosophy is often a limiting factor when the feasibility of the control system is evaluated. Management may limit the degree of automation because of the limited revenue and availability of technical expertise. Therefore, simple control methods may be favorable in lieu of more complex control methods. Proven control techniques and successful methods of operation may be favored over experimental techniques. Any of these limitations imposed by management need to be included in the feasibility evaluation.

4-16. Report

Usually, management decides whether or not to proceed with the implementation of a control system. Therefore, a clear, concise report is necessary to present the feasibility of the proposed control method. The feasibility presentation should be part of an overall planning report. The following suggests the basic needs of the report and summarizes the contents of this chapter.

- *Description* includes physical characteristics and objectives of the canal project.
- *Operating criteria* are established based upon project constraints and need to upgrade the canal operation.
- *Plan of operation* includes selection and application of operation concepts and methods of operation.
- *Application of control method* determines the practical control system that will satisfy the operating criteria and plan of operation.
- *Feasibility* estimates the total control system cost and value of tangible benefits. Intangible benefits are described and feasibility is evaluated based upon judgment and practical experience.
- *Selection* identifies the proposed control system and alternatives. The final control system selection may change during the procurement phase of the project.
- *Implementation* includes final design, writing specifications for equipment and software, and installing the control system on the canal project.

IMPLEMENTATION

4-17. Final Design

The implementation phase begins by establishing final design details for onsite and control equipment requirements. Mathematical model simulation studies may be required to finalize control software, such as local automatic control parameters. Considerable data gathered earlier in the planning process can be used for final design, but visits to the canal site may be required to acquire additional data.

4-18. Specifications

Specifications should be completed after the design is finalized. Writing a competent specification is critical to achieving a successful system. Incomplete or unclear specification requirements can lead to disastrous results. The completed specification text has many requirements that are necessary to achieve successful procurement of the desired control system.

A specification for control equipment should include the following major sections:

- *General requirements* include a description of canal project and control equipment, training, warranty, spare parts, tools, test equipment, and drawing and data requirements.
- *Materials and workmanship* include requirements for materials, quality of workmanship, reference specifications and standards, and work and material to be furnished under the specifications.
- *System functional requirements* include general data acquisition, data processing, electrical protection, transient and overvoltage protection, ambient environment, and equipment fabrication.
- *Hardware requirements* include detailed specifications for the control equipment.
- *Software requirements* include detailed specifications for the control software.
- *Testing, installation, and acceptance* includes requirements for factory acceptance testing, site testing, system performance acceptance testing, test equipment, installation, and maintenance.
- *Communication system requirements* include specifications for the required communication system, both hardware and software.
- *Training requirements* detail hardware and software operation and maintenance procedures.

Depending on the complexity of the equipment, the following requirements should be considered for inclusion in specifications:

- Inspections
- Deliveries
- Warranties
- Adjustments
- Socioeconomics
- Contract administration data
- Contract clauses
- Quality assurances
- Labor standards

Reference [9] is a Bureau of Reclamation specification for a supervisory control system.

4-19. Onsite Preparations

After the procurement phase, necessary onsite preparations can begin. For smaller canal projects it is often economical to have personnel from the operating entity do the work. The required onsite preparations must be completed before control equipment can be installed and field tested. To accomplish the work efficiently, it is important to have text and drawings that detail all the necessary requirements by location.

4-20. Testing and Installation

Testing the control equipment at the factory is of utmost importance. Before shipping to the site, factory tests are necessary to verify the control system will function according to the design. The testing program should include specified testing procedures with all control and communication equipment connected like it would be when installed on the canal project. Sensors should simulate actual field measurements so that the resulting control actions can be observed and checked for adequate performance. Usually, it is easier to debug equipment at the factory than in the field. Therefore, ensuring that all equipment functions satisfactorily before shipment simplifies the installation process.

The implementation phase concludes with the actual installation of the control system on the canal project. Frequently, control system installation is performed by the same contractor who furnishes the control equipment. However, installation contractors (specialists) perform this type work, and a separate contract (subcontract) may be issued. For smaller canal projects, often it is economical to have personnel from the operating entity install the equipment.

BIBLIOGRAPHY

4-21. Bibliography

[1] Sullivan, Edwin F., and Clark P. Buyalski, "Operations of Central Valley Project Canals," a paper presented at the ASCE Irrigation and Drainage Division Conference, Phoenix, Arizona, November 13-15, 1968.

[2] Buyalski, Clark P., and Edward A. Serfozo, "Electronic Filter Level Offset (EL-FLO) Plus RESET Equipment for Automatic Downstream Control of Canals," Bureau of Reclamation Report No. REC-ERC-79-3, 149 pp., Denver, Colorado, June, 1979.

[3] Buyalski, Clark P., "Automatic Downstream Control of the Corning Canal Check Gates by the EL-FLO Plus RESET Control System," a paper presented at the Symposium for U.S./U.S.S.R. Cooperative Program, *Automation and Remote Control of Water Resource Systems*, Frunze, Kirgiz, S.S.R., May 1979.

[4] ______________, "Basic Equipment in Automatic Delivery Systems," a paper presented at the National Irrigation Symposium: Irrigation Today and Tomorrow, Lincoln, Nebraska, November 10-13, 1970.

[5] ______________, "Navajo Indian Irrigation Project (NIIP)—Evaluation of Control Schemes," attachment to internal memorandum to Water Systems Automation Team Leader, code: D-430 from Chief, Division of Research and Laboratory Services, Bureau of Reclamation, Denver, Colorado, January 10, 1979.

[6] Graves, Albert L., and Robert S. Gooch, "Central Arizona Project Startup," a paper presented at the ASCE Water Forum '86: World Water Issues in Evolution, Long Beach, California, August 4-6, 1986.

[7] Abraham, Charles E., "Applied Hydro System Operations Models," a paper presented at the Water Forum '86: World Water Issues in Evolution, Long Beach, California, August 4-6, 1986.

[8] Falvey, H.T., and P. C. Luning, "Gate Stroking," Bureau of Reclamation Report No. REC-ERC-79-7, 145 pp., Denver, Colorado, July 1979.

[9] Solicitation/Specifications, 8-SP-40-03880/DS-7750, "Programmable Master Supervisory Control System, Dolores Project, Colorado," Upper Colorado Region, Bureau of Reclamation, Salt Lake City, Utah, 1988.

GLOSSARY

TERM	DEFINITION
A	
Actuator	Converts the output of the control element to a mechanical operation that effects the process operation.
Algorithm	A prescribed set of well defined rules or processes for the solution of a problem in a finite number of steps, usually expressed in the form of Boolean logic and mathematical equations.
Analog	A continuously variable electrical signal representing a measured quantity. For example, electrical signals such as current, voltage, frequency, or phase used to represent physical quantities such as water level, flow, and gate position.
Antihunt control	A stabilizing or equalizing system used to modify the response of a controller to prevent self-oscillations that would cause excessive operations of the controlled device.
Automatic control	A procedure or method used to regulate mechanical or electrical equipment without human observation, effort, or decision.
Automation	A procedure or method used to regulate a water system by mechanical or electronic equipment that takes the place of human observation, effort, and decision; the condition of being automatically controlled.
B	
Balanced operation	Operation of a canal system where the water supply exactly matches the total flow demand.
Boundary conditions	Flow conditions imposed at the ends of a pipeline or canal reach by various physical structures which must be described mathematically to solve the general equation of flow for hydraulic transient computer models.
C	
Canal automation	The implementation of a control system that upgrades the conventional method of canal system operation.
Canal check gate structure	A structure designed to control the water surface level and flow in a canal, maintaining a specified water depth or head on outlets or turnout structures. Most canal check structures have moveable gates.
Canal freeboard	The amount of canal lining available above maximum design water depth.
Canal pool	Canal section between check structures.
Canal prism	The cross sectional shape of a typical canal.
Canal reach	Segment of main canal system consisting of a series of canal pools between major flow control structures.

TERM	DEFINITION
Canal system control concepts	Downstream control and upstream control.
Canal system control methods	Local manual, local automatic, supervisory.
Canal system operation	Water transfer from its source to points of diversion for irrigation, municipal and industrial, fish and wildlife, and drainage purposes
Canal system operation concepts	Downstream operation and upstream operation.
Canal system operation methods	Constant downstream depth, constant upstream depth, constant volume, controlled volume.
Cascade control	A method of automation using control units arranged in sequence such that a given unit is controlled by the preceding unit and controls the following unit.
Cascade flow	Regulated flow through a series of flow control structures.
Centralized control	Control of a canal project from a central location generally by a master station, communications network, and one or more remote terminal units (RTUs).
Centralized headquarters	Control of a canal project from a central location by the watermaster.
Check gate	A gate located at a check structure used to control flow.
Closed conduit system	A conveyance system where the flow of water is confined on all boundaries; i.e., pipe systems.
Closed-loop control	A classification of control that corrects errors in the system being controlled by monitoring the controlled value, and comparing it with a standard representing the desired performance.
Collector system	Conveys water from several individual sources such as groundwater wells and drains and surface inlet drains for rain storm and snowmelt runoff to a single point of diversion. The collector system is associated with projects that increase water supply and decrease flood damage.
Colvin algorithm	A canal flow control structure technique that operates the gates based upon the rate of deviation of the water surface level from the setpoint.
Comparator	A circuit or device that compares two electrical signals and provides an indication of agreement or disagreement.
Connector system	Conveys water from a single source to a different location typically without intermediate collection or diversions. The connector system is associated with intakes to pumping plants and/or powerplants and with regulation reservoirs.
Constant head orifice turnout	A calibrated structure containing an adjustable orifice gate and a gate downstream to control a constant head differential across the orifice gate to divert and measure water from a main irrigation canal to a distributing canal.

TERM	DEFINITION
Constant volume operation method	A canal operation that maintains a relatively constant water volume in each canal pool.
Control	To exercise restraining or directing influence over; a mechanism used to regulate or govern operation of a system.
Control element	A part of a control system through which the system's process is regulated.
Control system	An arrangement of electronic, electrical, and mechanical components that commands or directs the regulation of a canal system.
Controlled variable	The quantity or condition of a system that is measured and controlled.
Controlled volume operation method	An operation in which the volume of water within a canal reach between two check structures is controlled in a prescribed manner for time variable inflows and outflows such as off-peak pumping or canal side deliveries.
Conventional method	Where operations personnel (ditchrider and watermaster) control the canal system onsite. Labor saving devices and machinery may be used to assist in the control of the canal facilities.

D

TERM	DEFINITION
Deadband	The range through which the measured signal can vary without initiating a control action.
Dead time	The time required for the response to a change of input to a system to reach the location of a sensor; i.e., the time for a control initiated surge wave to travel from an upstream control check gate to a downstream sensor in a canal.
Delivery system	Conveys water from a single source such as a storage reservoir to a number of individual points of use. The delivery system is a common classification. It is associated with irrigation, municipal and industrial, and fish and wildlife canal systems.
Demand delivery	Unrestricted use of the available water supply with limitations only on maximum flow rate and total allotment.
Derivative (rate) control	A mode of control that proportions the control device setting to the rate of change of the controlled variable. Derivative control is usually used with proportional or proportional plus integral control.
Digital	Representation of a quantity by an arrangement of digits each of which represents a portion or weighted portion of the quantity. In communications, transmission of quantities by a series of pulses.
Distribution system	Delivers water from the main canalside turnout to individual water users or to other smaller distribution systems.

TERM	DEFINITION
Ditchrider	Canal system operations personnel. The person responsible for controlling the canal system onsite based upon the flow schedule established by the watermaster.
Diversion dam	The diversion dam is commonly constructed on a natural river channel and designed to check or to elevate the water level for diversion into a main canal system.
Downstream control	Control structure adjustments are based upon information from downstream. The required information is measured by a sensor located downstream or based upon the downstream water schedule established by the watermaster.
E	
EL-FLO controller	The EL-FLO (electronic filter level offset) controller is a proportional and integral controller that utilizes an electronically filtered (delayed and smoothed) water level signal from the downstream end of a canal reach. EL-FLO controllers are used for control of multiple reach canals and require communications circuits between check structures.
F	
Feedback	The application of a portion of a signal or a function of a signal to a preceding stage of a control system.
Feed forward	The application of a portion of a signal or a function of a signal to a forward stage of a control system.
Filter	A device used to remove specific unwanted frequencies from electrical, mechanical or hydraulic systems.
Floating control	See three position control.
Frequency response analysis	A method of studying systems in which sine wave disturbances are input to the system and the corresponding amplitudes and phases of these sine waves in the output signal are determined.
G	
Gain	The ratio of the output to the input signal of a control system. Gain is used to describe the proportionality factor of a proportional control system.
Gate position sensor	A device such as an analog or digital sensor that can measure the mechanical position of a gate and provide a signal representing the position.
Gate stroking	A predetermined method of controlling variation in gate speed in open channels to regulate the magnitude of water surface elevation disturbances as discharge is varied.

TERM	DEFINITION
	H
Hardware	Physical components of control systems. For example, relays, switches, lights, integrated circuits, transistors, and capacitors are hardware components.
Hydraulic gradient pivot point	A location along the water surface in a canal reach where the water level remains essentially constant during changes in flow.
Hydraulic transient	A wave or pressure change propagated through a canal or pipeline during unsteady flow.
	I
In-channel storage	See inline reservoir.
Inline reservoir	Constructed inline with the canal and consists of a large pool of the main canal. In canal systems, used to regulate flow for balanced operation.
Instability	A condition of a system in which the state of equilibrium between elements is not restored.
Integral (reset) control	A mode of control in which the output is proportional to the time integral of the error. Integral control is used in conjunction with proportional control in a hydraulic system.
Interface	A combination of electrical, mechanical, or other components interconnecting elements of a control system.
Inverted siphon	A closed pipe used to convey the main canal under drainage channels, depressions, roadway, or other structures. Also referred to as a sag pipe.
	L
Lateral	A branch in a canal or pipeline system that diverges from the main canal or other branches.
Linearization	Replacement of a nonlinear equation by an approximate linear equation or equation.
Little-Man controller	A variety of canal controllers of the floating control type that utilize a set operate time and a set rest time mode of control and usually include an antihunt device.
Local automatic control	Onsite control by control equipment without human intervention.
Local manual control	Onsite control by a human operator (ditchrider).
	M
Main canal system	Delivers water from a primary source of supply to several points of diversion or canalside turnouts to smaller distribution systems.

TERM	DEFINITION
Manual control	Control of equipment requiring direct intervention of a human operator.
Master station	The centralized facility with communications to remote terminal units for the purpose of information retrieval, control of apparatus, system control, and operation optimization.
Mathematical model	A representation of physical laws or processes expressed in terms of mathematical symbols and expressions. The model is used as a basis for computer programs for examining the effect of changing certain variables in the analysis of the effect of flow changes in a water delivery system.
Microwave radio	A method of point-to-point radio transmission using the frequency spectrum above 890 MHz. Characteristic of microwave frequencies limit transmission to line of sight distances.
Mismatch	A condition in which water supplied to a given point in a conveyance or distribution system does not equal the demand for water at that point.

O

TERM	DEFINITION
Offline reservoir	Constructed to the side of the main canal usually in a natural drainage channel. In canal systems, used to store surplus water runoff during the winter season for use during the irrigation season.
Offset	The difference between the controlled variable and the referenced input; i.e., the difference between the water level in a canal system and the water level at design flow.
On-Off control	See two position control.
Open channel system	Where the top flow boundary is a free surface; i.e., canal systems.
Open-loop control	A classification of control that initiates a control action with no comparisons to actual process conditions.
Operational spill	A loss or waste of water in an irrigation system caused by operation of the system.
Overshoot	The maximum difference between the transient and steady-state controlled variable (water level or discharge). A measure of relative stability represented as a percentage of the final value of the output (steady-state condition).

P

TERM	DEFINITION
P+PR	P+PR (proportional plus proportional reset) algorithm is similar to the EL-FLO + Reset algorithm except that the P+PR algorithm is applied to the automatic upstream control of canal systems.
PID controller	A controller that combines proportional, integral, and derivative modes of control.

TERM	DEFINITION
Program	A set of coded instructions that direct a computer to perform some specific function or yield the solution to some specific problem.
Proportional control	A mode of control that moves a controlled element to a position proportional to the difference (offset) between the actual value of the controlled variable and the target.
Proportional plus reset control	A combination of proportional and integral control in which the difference or offset caused by the proportional mode of control is eliminated by the reset action.
Pumped storage	A reservoir that has a pumping plant as the main source of water supply. Also referred to as an offline reservoir.
R	
Regulation reservoir	A reservoir used in canal systems to reduce the mismatch between downstream demands and upstream water supplies to maintain a balanced operation.
Remote monitoring	Periodic or continuous measuring of quantities at remote sites for transmission and dissemination at another location.
Remote terminal unit (RTU)	Supervisory control equipment at the remote site that performs data collection, executes control commands, performs automatic control functions, and communicates with a master station.
Response time	The time required for the depth, pressure, or flow to reach and remain within a certain percentage of its steady-state value after a control correction has been initiated.
Rotation delivery	Water delivery where a relatively constant supply flow is rotated to different users at varying times.
S	
Sag pipe	See inverted siphon.
Scheduled delivery	Operation of a water delivery system to meet predetermined needs, generally based upon user water orders.
Self regulation	A controlled system requiring virtually no operator intervention (see automatic control).
Sensor	A device for measuring water level, flow, gate position, etc., for input to a local automatic controller or RTU.
Sequential control	Similar to setpoint positioning except allows a single operator command to execute a control sequence involving several steps.
Setpoint	A value of water level, flow, etc., that the control system maintains, also called the target.

TERM	DEFINITION
Setpoint positioning	A control operation that compares an existing value, such as water level, with a desired value and uses the difference (error) to initiate a corrective action (raise or lower a gate; start or stop a pump).
Software	Coded instructions that direct the operation of a computer. A set of such instructions for accomplishing a certain task is called a program.
Stable canal system	A canal system in which flow disturbances are attenuated.
Steady flow	Flow which is constant with respect to time.
Storage reservoir	Collects and stores water from storm runoff and snowmelt. It is the primary source of supply to the water project and main canal system.
Supervisory control	The control of a canal system from a centralized location (master station) over a communication system and using remote terminal units (RTUs) at the canal structure sites.
Surge wave	A translatory wave in an open channel resulting from a sudden change in flow of water, such as that caused by opening or closing a gate.

T

TERM	DEFINITION
Target	A value of water level, flow, etc., that a control system maintains, also called the setpoint.
Telemeter	To sense, encode, and transmit data to a distant point.
Three-position control	A mode of control that responds to the deviation from the setpoint by operating the controlled device for a predetermined amount of time. The corrective signal has three discrete values: no correction; increase the output correction; and decrease the output correction.
Time lag	The time required for a change in the input to a system to cause a change in the output. Time lag is a result of dead time, measurement delay between sensing point and controller, or delay in signal transmission between controller and process.
Transducer	A device that converts one form of energy to another, such as hydraulic to pneumatic, or mechanical to electrical.
Transient flow	Unsteady flow during a change from a steady-flow state to another steady-flow state.
Transient response analysis	A method of analysis in which a system is subjected to a pulse input and the output of the system is recorded and analyzed with respect to fluctuations in flow with time.
Translatory wave	A gravity wave that propagates in an open channel and results in displacement of water particles in a direction parallel to the flow.

TERM	DEFINITION
Turnout	A structure provided to divert water from a main or primary irrigation canal to a distribution canal or farm delivery point. Turnouts are used at the head of canal laterals.
Two-position control	A mode of control that responds to the deviation from setpoint by operating the controlled device to either of two extreme positions. Also referred to as On-Off control.

U

TERM	DEFINITION
Unsteady flow	If at a specified location along a water conveyance, there is a change in flow with respect to time, the flow is unsteady at that location.
Upgrade	Provides a better match between the canal system delivery capabilities and the water users demands. As a result, improved response and efficiency of a system is achieved beyond what could be accomplished by the conventional method of operation.
Upstream control	Control structure adjustments based upon information from upstream. The required information is measured by a sensor located upstream, or based upon the upstream water schedule established by the watermaster.

V

TERM	DEFINITION
Variable target	A control technique that uses a changing setpoint value or target to compensate for changes in demand or storage.

W

TERM	DEFINITION
Wasteway	Structure used to divert surplus flow from the main canal into a natural or constructed drainage channel.
Water level pivot point	A location along the water surface in a canal reach where the water level remains essentially constant during changes in flow.
Watermaster	The person responsible for operation of the entire canal project.
Wave celerity	The velocity of propagation of a wave through a liquid, relative to the rate of movement of the liquid through which the disturbance is propagated.
Wedge storage	The volume of water contained between two different water surface profiles (flow changes) within a canal pool.

TERM	DEFINITION
Turnout	A structure provided to divert water from a main or primary irrigation canal to a distribution canal or farm delivery point. Turnouts are used at the head of canal laterals.
Two-position control	A mode of control that responds to the deviation from setpoint by operating the controlled device to either of two extreme positions. Also referred to as On-Off control.
	U
Unsteady flow	If at a specified location along a water conveyance, there is a change in flow with respect to time, the flow is unsteady at that location.
Upgrade	Provides a better match between the canal system delivery capabilities and the water users demands. As a result, improved response and efficiency of a system is achieved beyond what could be accomplished by the conventional method of operation.
Upstream control	Control structure adjustments based upon information from upstream. The required information is measured by a sensor located upstream, or based upon the upstream water schedule established by the watermaster.
	V
Variable target	A control technique that uses a changing setpoint value or target to compensate for changes in demand or storage.
	W
Wasteway	Structure used to divert surplus flow from the main canal into a natural or constructed drainage channel.
Water level pivot point	A location along the water surface in a canal reach where the water level remains essentially constant during changes in flow.
Watermaster	The person responsible for operation of the entire canal project.
Wave celerity	The velocity of propagation of a wave through a liquid, relative to the rate of movement of the liquid through which the disturbance is propagated.
Wedge storage	The volume of water contained between two different water surface profiles (flow changes) within a canal pool.